POCKET GUIDE TO
TREES
AND
SHRUBS

First published in 2014 by Bloomsbury Publishing Plc,
50 Bedford Square, London WC1B 3DP

www.bloomsbury.com

ISBN (print) 978-1-4729-0981-7

Bloomsbury Publishing, London, New Delhi, New York and Sydney

Bloomsbury is a trademark of Bloomsbury Publishing Plc

A CIP catalogue record for this book is available from the British Library
Library of Congress Cataloging-in-Publication Data has been applied for

Publisher: Nigel Redman
Project editor: Jane Lawes
Design by Rod Teasdale

Printed in China by Toppan Leefung Printing Co Ltd.

This book is produced using paper that is made from wood grown in managed
sustainable forests. It is natural, renewable and recyclable. The logging and
manufacturing processes conform to the environmental regulation of the
country of origin.

10 9 8 7 6 5 4 3 2 1

POCKET GUIDE TO
TREES
AND
SHRUBS

Bob Gibbons

BLOOMSBURY
LONDON · NEW DELHI · NEW YORK · SYDNEY

CONTENTS

The female flowers of Hazel.

Old Beech pollards, Epping Forest.

INTRODUCTION

WHAT ARE TREES AND SHRUBS?

Most people are familiar with the general features of trees and shrubs, though it is not always easy to define exactly what is meant by a 'tree' or 'shrub'. Both are characterised by having woody stems and branches; this is achieved by a process of secondary thickening, where the stems increase in girth each year by laying down woody cells within the tissues, produced by special layers. The diameter of the trunks and branches can continue to increase throughout the life of a tree.

In general, trees are thought of as woody plants that have a main stem that is 5m (16ft) or more in height, with a branching crown above this, whereas shrubs have numerous branches arising lower down and are usually less than 5m (16ft) in height. There is, of course, a good deal of overlap between these two categories, particularly with respect to young trees, trees growing in challenging conditions (such as high in the mountains, or in the far north) where they remain small, or shrubs growing in especially favourable conditions, where they grow large. Some plants, such as Common Juniper, Hawthorn or Hazel, are most commonly shrubs, but can certainly become small trees under the right conditions. In this book, we have included both trees and shrubs, so the distinction is less significant.

We have included a few shrubs that are very small, often under 50cm (1½ft) in height, which might not normally be thought of as shrubs at all, such as Dwarf Birch, Bog Myrtle or Net-leaved Willow; however, these are fully formed shrubs, with woody stems, often closely related to much more typical shrubs or trees, and it makes sense to include them.

THE SCOPE OF THIS BOOK

This book covers the great majority of native trees and shrubs of Britain and north-west Europe, excluding only some of the more difficult species from large groups such as Willows *Salix* spp. In addition, it covers a number of the most common, conspicuous or distinctive non-native species that may be found planted or naturalised in the area. Some of these, such as the Giant Redwood or the Cedar of Lebanon, will only be found as planted specimens, as they have little ability to invade their surroundings; others, such as Sycamore or Rhododendron, have become so much a part of our vegetation that they are often simply thought of as being native.

Hawthorn In flower in May.

HOW TO USE THE BOOK

Under each species, there is a general description of the tree or shrub, with an indication of its average maximum height. Occasional individuals will be taller than this, and of course many young or poorly grown examples will be smaller – it is just a guide to how large the tree normally becomes. Within this description, key features that help to identify the species are picked out, and – in some cases – specific reference to confusable species is made, with a note of the differences. Normally, details of the flowers, and the subsequent fruit if distinctive, are given.

Many trees have separate male and female plants, e.g. Holly, or separate male and female flowers on the same tree, e.g. Hazel, where the familiar catkins are the male flowers, while the female flowers are small and inconspicuous. These differences are normally described.

The heading '**Habitat and distribution**' gives a description of what types of places the tree is most likely to be found in, e.g. mountain woodlands, river floodplains, etc., and whether or not it is likely to be planted or naturalised. Some species are only found as natives outside Europe, but have been planted here; for these, the country of origin is normally noted. Other species are native in this area, but have also been widely planted – e.g. Wild Pear – and an indication is given, if it is known, as to which is the original native area.

'**Flowering time**' gives an indication of the most likely time of flowering. However, in a large region such as north Europe, flowering times vary considerably with latitude, and also with altitude wherever there are hills or mountains, so this is just a guide. They will generally be later at higher altitudes or further north. Also, trees will not flower at consistently the same time each year, and a general trend for earlier flowering has been observed in many species, presumably due to global warming.

In some instances, the heading '**Similar species**' gives a briefer description of any additional similar or closely related trees and shrubs that also occur in the region. For example, Common Laburnum is an easily recognisable native or garden plant in much of Europe; but there are also some other very similar Laburnum species that could be confused with it, and these are described in this section.

There are a few groups of trees and shrubs, such as the roses *Rosa* spp., the elms *Ulmus* spp. and to a lesser extent the willows

Hips and leaves of Dog-rose.

Salix spp., that consist of a confusing mixture of species, hybrids, varieties and microspecies, often still subject to discussion amongst taxonomists. These have been treated slightly differently, more as a general description of the group, with some of the more obvious species picked out.

A number of specialist terms used in the text are defined in the **glossary**, rather than defining them each time they are used.

Because there are relatively few species of trees in this region, the book is most easily used by simply scanning through the photographs to check for anything looking like your specimen. The description should assist with refining your identification and, where relevant, the '**Similar species**' section should be checked to see if any other species makes a better match. If the tree is wild or naturalised, it should be possible to identify it with a reasonable degree of certainty; if it is a specimen tree, such as in a park or garden, it may be a species not covered in the book (there are thousands of different trees and shrubs from around the world planted in north Europe), and it may only be possible to make a rough guess.

HOW TO IDENTIFY TREES

If you are especially interested in identifying trees, there are a few items of equipment that are worth purchasing and carrying with you to assist in the process. A small hand lens, of about 8x magnification, is easily carried and will be enormously helpful in allowing a closer examination of the structure, shape or hairiness of the parts of a plant. It also opens up a fascinating new world. These lenses are available from natural history suppliers, jewellers or, occasionally, opticians. A small pocket notebook, with pen or pencil, is ideal for noting down the shape and height of a tree, where it was, what photos were taken, etc., or the details from a label if in a garden. A small ruler is also useful to check the length of leaves, etc.

A camera – especially a small, compact, modern digital camera – is an ideal addition to the kit, for photos of whole trees or close-ups of leaves, details, fallen fruit, etc. (and these can often reveal additional detail, not seen by the eye, later when enlarged at home), and also of labels in gardens for identification. Incidentally, it is a good idea to get into the habit of consistently photographing the label straight *after* you photograph the plant, so that you later know which name refers to which specimen, especially if you are photographing several.

When studying an unknown specimen, it's worth looking at the general shape, especially of a tree (e.g. is it narrowly columnar, spreading, flat-topped, multi-branched, heavy-limbed, and so on). Leaf shape can be crucial, noting especially the general shape of the leaf blade, how it joins to the leafstalk, whether it is pointed or rounded at the tip, whether the margin is toothed or plain, and how long the leafstalk is. For some species, especially among the conifers, it may be important to note how the leaves are arranged (e.g. are they all around the stem, or flattened into two rows).

The details of flowers or fruit can often make an identification more accurate or easier. Check carefully to see if there are any flowers or fruits on the tree, looking particularly on higher branches and on the south side if none are immediately visible. Sometimes, there may be a fallen example on the floor, which can be collected for later examination. If possible, apart from obvious features such as colour and size, check the flowers for number of petals, and whether they are male only, female only or both. Fruit size and colour should be examined, and with conifer cones it's worth noting how the cones are held (e.g. erect or hanging). A small pair of binoculars can be useful for checking out-of-reach flowers or fruits, but it's also possible to take a photograph, at the telephoto setting if there is one, then examine the enlarged photos later for more details.

ESTIMATING THE HEIGHT OF TREES
Although not essential for identification, it can be useful and interesting to know the approximate height of a large tree, and there are several simple ways of discovering this.

If there are two of you, ask your companion to stand at the base of a tree. Find (or break) a branch of roughly the same length as your arm, hold it vertically, then walk away from the tree until the length of the stick, when at arm's length, appears to be the same height as the tree. Turn the stick through 90°, either to left or right according to which is easier to walk, and ask your helper to walk along the ground until they reach the apparent end of the stick. This will be the same height as the tree.

Alternatively, follow the same procedure on your own, and – after finding the position where the top and bottom of the stick match the top and bottom of the tree when held at arm's length – walk from that position to the base of the tree. This will be the approximate height of the tree.

Coast Redwood trees.

Silver Fir
Abies alba

A tall, rather elegant, pyramidal to conical evergreen tree, up to 50m high with smooth greyish-brown bark that eventually becomes fissured. The leaves (needles) are short, up to 3cm long, leathery, blunt and notched at the tip. They are coloured green above but with two prominent white bands below. Though borne all round the stem, the leaves spread out laterally to give a 'parting' along the top of the twig. The female cones are erect (the firs can be distinguished from spruces by the erect cones), oblong, woody and up to 20cm tall, with a projecting spine between each scale. The scales and bracts are deciduous, leaving the central peg-like axis behind when they fall.

 FLOWERING TIME March–April.

HABITAT AND DISTRIBUTION Silver Fir is native to middle-altitude forests in the Alps and surrounding mountain areas, but it is also very widely planted as an ornamental or forestry tree almost throughout the area, and occasionally naturalised.

SIMILAR SPECIES Giant Fir *A. grandis* can be up to 60m high, with brighter green foliage, and cones with the bracts enclosed, not readily visible. From western North America, they are widely planted for forestry or ornament, and occasionally naturalised.

Above: Inside Silver Fir woodland.

Right: A mature Silver Fir tree.

Douglas Fir or Oregon Fir
Pseudotsuga menziesii

A very tall, slender, conical tree, up to 65m tall (but reaching 100m – and formerly much taller – in its native North America), with grey-green or reddish bark becoming corky and fissured with age. The branches are in whorls, level or drooping, with masses of dark pendulous foliage. The leaves are narrow, up to 3.5cm long, blunt, aromatic, dark green above but with 2 prominent white lines underneath. The female cones are narrowly oval, pendent (unlike those of true firs, *Abies*, which are erect), up to 10cm long and distinctive by virtue of the strongly 3-toothed bracts – with a longer central bract – on each cone scale. These cones fall complete – unlike those of firs, which split up. The male cones are small, yellow, and in the axils of the old shoot leaves.

FLOWERING TIME April–May.

HABITAT AND DISTRIBUTION It is native to western North America, but is very widely planted, especially in Scotland, as a fast-growing forestry and parkland tree almost throughout the area (though much more rarely in Ireland), and occasionally naturalised.

Above: The distinctive three-toothed bracts of the cones.
Left: A planted Douglas Fir.
Right: A very large Douglas Fir in its native Oregon.

Norway Spruce
Picea abies

A tall, conical tree, up to about 60m high, with whorled branches, roughly level, though lower branches are often descending in older trees and frequently reaching the ground. It is well known in the UK as the Christmas tree. The bark is reddish-brown and scaly. The needles are stiff, short (up to 2.5cm long), pointed, dark green and 4-angled in cross section, each borne on a short woody peg, characteristic of the spruces. The male cones are small, ovoid, yellow or reddish and clustered near the tips. The female cones are narrowly ovoid, pendent, up to 18cm long and without bracts between the scales.

 FLOWERING TIME March–April.

HABITAT AND DISTRIBUTION It is native throughout most of northern Europe in mountain and boreal areas (though not native to Britain), but very widely planted as a forestry, shelter-belt and ornamental tree almost throughout the area, and occasionally naturalised. It is widely grown as a young tree for use in the Christmas-tree market.

SIMILAR SPECIES There are no other native spruces in northern Europe, though many exotic species are planted. The most widespread is Sitka Spruce *P. sitchensis*, from western North America, which has bluish-green foliage and purplish bark.

Left: Pendulous female cones of Norway Spruce.

Right: A mature Norway Spruce in winter.

European Larch or Common Larch
Larix decidua

A medium-sized, deciduous, coniferous tree, narrowly conical in shape and up to 45m high. The bark is rough, greyish-brown and becomes cracked and fissured as the tree matures. The branches are rather short, level or descending. The twigs are yellowish, bearing short shoots with clusters of 30 to 40 needles, as well as scattered needles elsewhere on the shoot. The needles are up to 3cm long, fresh green when young, with two inconspicuous greenish stripes on the underside. In autumn they turn golden yellow or reddish before falling, producing a spectacular display. The female cones are bright red when young, becoming ovoid, brown and woody, about 3cm long, as they age, with erect unreflexed cone scales. The male flowers occur in small, pale yellow cones, releasing pollen in spring.

 FLOWERING TIME The flowers appear in March–April, but the cones persist throughout the year.

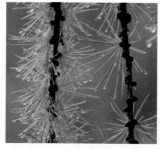

Typical clusters of larch needles.

Mature female cones.

HABITAT AND DISTRIBUTION It is native to middle- or higher-altitude forests in the Alps, but it is also very widely planted as an ornamental or forestry tree almost throughout the area, and occasionally naturalised.

SIMILAR SPECIES Japanese larch *L. x kaempferi* is similar, but it has bolder white stripes on the leaves, and the female cone scales are distinctly turned back. It is native to Japan, but commonly planted for timber.

Right: Old Common Larch trees in autumn.

Scots Pine
Pinus sylvestris

A medium-sized tree, up to 35m tall. As it matures, it becomes distinctive, with a long branchless trunk, domed open crown, and red-brown scaly or papery bark, particularly in the upper parts of the trunk. The needles are borne in pairs (with pines, it is important to look at the number of needles in each bundle, which may be 1, 2, 3, or 5), and are greyish-green, quite long, up to 8cm, and twisted. The male flowers are yellow, borne in clusters towards the ends of the branches, shedding large quantities of yellowish pollen in spring. The female cones are solitary or paired, beginning as soft crimson clusters, gradually becoming brown and woody then ripening fully in the second year to a woody oval-conical cone up to 8cm long. Each cone scale has a small prickle in the centre.

Right: Male flowers.

Female cone with male flowers.

An ancient native Scots Pine in Scotland.

 **FLOWERING TIME** March–May.

HABITAT AND DISTRIBUTION Scots Pine is native, or formerly native, throughout most of northern Europe in a variety of habitats and soil types (though now confined in Britain, as a native tree, to the highlands of Scotland). It is very widely planted as a forestry, shelter-belt and ornamental tree almost throughout the area, and frequently naturalised.

21

Maritime Pine
Pinus pinaster

A medium to large tree, often with a largely bare trunk below when mature, with fissured reddish-brown bark, topped by a rounded crown. The needles are in pairs, long, up to 25cm, rather stiff and leathery, greyish-green and pointed. The male flowers are yellow, ovoid and in clusters near branch tips; the female cones are red at first, becoming shiny, brown, ovoid, large, up to 22cm long, and remaining on the branch after the leaves have fallen. It is best distinguished from Scots Pine by the long, leathery needles and large cones.

 FLOWERING TIME March–May.

HABITAT AND DISTRIBUTION It is a native of western Mediterranean coastal areas, but widely planted for timber, shelter and dune stabilisation, being particularly suited to sandy soils and coastal areas.

Mature female cone of Maritime Pine.

Maritime Pine needles and female cones.

SIMILAR SPECIES Corsican Pine *P. nigra* ssp. *laricio* is large and up to 50m tall. The leaves are in pairs, soft, narrow, up to 15cm long and pale green. The cones are small and ovoid. It is native to Corsica and adjacent areas, but widely planted for timber, shelter-belts and stabilisation, especially in coastal areas. Dwarf Mountain-pine *P. mugo* is much smaller, rarely over 4m, with twisted branches and short, curved, rigid needles. It is native to the mountain areas of central and southern Europe, but widely planted for ornament, shelter and stabilisation.

Dwarf Mountain-pine needles and female cones.

Cedar of Lebanon
Cedrus libani

A medium-sized conical tree, up to 40m high, becoming flat-topped at maturity, with a thick, solid trunk. The bark is greyish brown, ridged and fissured. The main branches are usually massive, ascending and ending in a series of level, lateral branches with plateaux of dense needles. The needles are borne in clusters of 10–15 on short shoots and are each 2–3cm long and dark green in colour. The male cones are slender, greyish-green, up to 7cm long and borne in clusters at the end of shoots. The female cones of cedars are distinctive, barrel-shaped or conical, solid, erect and ripening from blue-green to woody brown.

 FLOWERING TIME September–October.

Cedar of Lebanon forest.

HABITAT AND DISTRIBUTION It is native to middle- or higher-altitude forests in the eastern Mediterranean, but very widely planted as an ornamental tree almost throughout the area, becoming a feature of old parks and gardens.

SIMILAR SPECIES There are two other main species of cedars (and one minor one), all rather similar. A useful way of distinguishing mature trees is by their terminal branches, which are roughly Level in Lebanon, Ascending in Atlas and Descending in Deodar. Atlas Cedar *C. atlantica*, from north-west Africa, has needles in clusters of 30 or more and is widely planted, especially as 'glauca', which has grey-blue foliage. Deodar Cedar *C. deodara* from the Himalayas has 15–20 leaves per cluster. Cyprus Cedar *C. libani* ssp. *brevifolia* has shorter needles and is endemic to Cyprus, but occasionally planted elsewhere.

Female cone of Cedar of Lebanon.

A mature Atlas Cedar.

Monkey Puzzle or Chilean Pine
Araucaria araucana

A medium-sized, conical or domed evergreen tree, up to 30m tall. The trunk is cylindrical, straight, with greyish, wrinkled bark. The branches are densely covered with spirally arranged, scale-like ovate leaves, up to 4cm long and 2cm wide, with a sharp, brownish point. The male cones are narrow, up to 10cm long, and borne in clusters towards the branch tips. The female cones, on separate trees (i.e. the plant is dioecious), are globose, erect and up to 15cm across.

 FLOWERING TIME The flowers open in June–July.

HABITAT AND DISTRIBUTION It is native to middle- or higher-altitude forests in the Andes, but is very widely planted as an ornamental tree almost throughout the area, and occasionally naturalised.

A mature Monkey Puzzle tree.

Female cones and foliage.

Monterey Cypress or Macrocarpa
Cupressus macrocarpa

A large evergreen tree, up to 40m high, that is conical when young but becomes broadly domed with heavy branches, ascending at first but gradually flattening with age. The leaves are small, scale-like, 1–2mm long and borne on stiff, forward-pointing shoots, with a resinous lemon smell. Male cones are small, yellow and on shoot tips below the female cones, which are roughly spherical, up to 4cm in diameter, leathery and covered by 10–12 scales each with a central protuberance.

 FLOWERING TIME June.

HABITAT AND DISTRIBUTION Native to a small area of coastal California, but widely planted in western Europe for ornament and shelter, especially in coastal western regions, where there are some very large trees.

Left: Ripe female cones, and foliage, of Monterey Cypress.

Typical mature Monterey Cypress tree.

Wellingtonia or Giant Redwood
Sequoiadendron giganteum

An enormous and distinctive evergreen tree, reaching over 90m in the wild in California, or over 50m in Europe, with a narrow conical shape and a huge tapering trunk up to 7m or more across at the base. The bark is red or brown, spongy, deeply ridged and furrowed, shredding to reveal lower layers. The lower part of the trunk is usually bare, with the first branches well above head height, drooping but ascending at the tips. The leaves are small, less than 1cm long, and scale-like, arranged tightly around the shoots. The male cones are small, yellowish and at the ends of branches. The female cones are solitary or paired, ovoid in shape, up to 8cm long (usually smaller), woody and brown.

 FLOWERING TIME March–April.

HABITAT AND DISTRIBUTION It is native to a small area of inland California, where it was discovered in 1852, but widely planted in western Europe for ornament in parks and large gardens, especially in western regions (except Ireland), where there are some very large trees.

SIMILAR SPECIES The Coast Redwood *Sequoia sempervirens* is rather similar in size and shape, but differs in having two types of leaves – rather similar scale leaves, but also flattened spreading needle leaves, up to 2cm long, held in 2 rows. The female cones are smaller, up to 2.5cm, and more open. It is less common in cultivation.

Left: A grove of ancient Giant Redwoods.
Right: Giant Redwood trees at night.

Yew
Taxus baccata

A large bush or medium-sized tree, up to 28m high, sometimes with multiple trunks. The trunk is broad, irregular, often twisted, with reddish bark often peeling in large patches. The leaves are needle-like, narrow and flattened, dark green and shiny above, but paler below with two pale bands, and are up to 4cm long and borne spirally but usually flattened and twisted into two rows. The male flowers are solitary clusters of yellowish anthers at the base of leaves. The female flowers (on different trees) are tiny, but develop into distinctive bright red, ovoid, fleshy fruits, flattened at the end – the so-called berries.

 FLOWERING TIME March–April, and the distinctive fruits ripen in autumn.

HABITAT AND DISTRIBUTION It is native throughout much of Europe, especially, but not exclusively, on calcareous soils in a variety of habitats, occasionally forming forests. It is also widely planted, especially in churchyards in Britain, where the trees may be exceptionally old. Most parts of the tree are highly poisonous.

SIMILAR SPECIES Although there are a few other species of Yew, none are native or widely planted.

Male flowers of Yew.

Ripe Yew berries.

Common Juniper
Juniperus communis

A small evergreen tree or large shrub, up to 7m tall and occasionally more, that is normally rounded or conical in shape, but natural columnar forms occur in some areas. The bark is reddish-brown and peels off in papery patches. Twigs are stout, 3-angled and bearing whorls of 3 linear, stiff, pointed leaves up to 2.5mm long but only 1–2mm wide. Each leaf is dark green and keeled below, with a single broad, whitish band above (at times it may look like two bands due to the presence of a midrib vein), and smells fragrant when crushed. Male cones are solitary, yellow and produced in the leaf axils. The female cones (normally on separate trees) are the familiar juniper 'berries', roughly spherical, up to 1cm across, and bluish-black in colour. The berries are widely used as a flavour for gin and as a culinary spice.

 FLOWERING TIME March–April.

HABITAT AND DISTRIBUTION Native throughout most of Europe and in many other parts of the world, in a variety of habitats. There are distinctive dwarf subspecies found in western coastal areas or in mountains.

Juniper bushes in limestone grassland.

Ripening Juniper berries.

Barberry
Berberis vulgaris

A densely branched, spiny, deciduous shrub, up to about 3m high, with ridged yellow-brown twigs. The spines are 3-pointed and scattered along the stems at the nodes. The leaves are elliptic to ovate, up to 6cm long, undivided, but edged with fine, spiny teeth. The flowers are bright yellow, cup-shaped, about 6–8mm across and borne in tight pendulous racemes. The berries are narrowly oblong and bright, shiny red when ripe, up to 1.5cm long, edible and rich in vitamin C, but too acid for some tastes.

 FLOWERING TIME May–June.

HABITAT AND DISTRIBUTION It is native throughout much of Europe (though almost certainly only present as an ancient introduction in Britain and the more northern parts of Europe) in hedges, scrub and woodland edges.

SIMILAR SPECIES There are no other native species, but a number of exotics are cultivated and occasionally naturalised, notably Darwin's Barberry *B. darwinii*, with orange flowers and blue-purple fruits, from South America. The closely related *Mahonia* spp., such as Oregon Grape *M. aquifolium,* have pinnate leaves and no stem spines.

Barberry flowers in spring.

Barberry berries.

London Plane
Platanus x hispanica (= P. x hybrida)

A large, irregularly domed, deciduous tree, with heavy, spreading branches, up to about 45m high. The bark is distinctive, grey or brown, with large sections peeling away to reveal yellowish or paler patches below. The leaves are large, up to 18cm long, shallowly and palmately lobed into 3 or 5 sharply toothed lobes, with a hollow base to the petiole, concealing the bud. Male flowers are in pendent clusters of 2–6 yellowish, spherical heads. The female flowers are in 2–5 reddish, globose heads, ripening to spherical brown fruits, from which abundant hairy seeds disperse.

Fruits of London Plane in autumn.

FLOWERING TIME May–June.

HABITAT AND DISTRIBUTION The London Plane is of uncertain origin and not known as a native anywhere, but very widely planted, especially as a street tree. It is very tolerant of pollution and well suited to use in cities, where it may often be heavily pollarded.

SIMILAR SPECIES Oriental Plane *Platanus orientalis* is similar, but has more deeply divided leaves. It is native to south-east Europe, sometimes planted, and very occasionally naturalised.

Fallen London Plane leaves.

Box

Buxus sempervirens

An evergreen shrub or small tree, normally up to about 5m tall
and occasionally much taller, with yellowish-grey bark. The leaves
are small, up to 2.5cm long, leathery, oval, notched at the tip,
untoothed, dark green and in opposite pairs. The flowers are tiny,
just 2–3mm across, without petals, and are found in small
yellowish clusters. The male and female flowers are separate but
on the same plant (a condition known as monoecious). The fruit is
a distinctive, small, woody, green – then brown – 3-horned
capsule up to 1cm long that eventually splits to scatter the hard,
shiny, brown seeds. The hard, fine-grained wood has a variety of
small-scale uses, such as for cabinets or musical instruments.

 FLOWERING TIME April–May.

HABITAT AND DISTRIBUTION It is native throughout southern
and central Europe, as far north as Belgium and central Germany,
in scrub, wood margins and other open habitats on drier soils. It is
planted and naturalised further north and is rare on calcareous
soils in southern Britain as a native, but widely planted elsewhere.

Box flowers and developing 3-horned fruits.

Box in flower.

Redcurrant
Ribes rubrum

A small, erect, deciduous much-branched shrub, up to 2m high. The leaves are up to 10cm long, palmately lobed into 5–7 triangular lobes, more or less hairless, and unscented. The flowers are pale green, sometimes tinged pinkish, 5-parted, each about 5mm across, and in a drooping raceme of up to 20 flowers. The fruit is the familiar redcurrant, globular, bright, shiny, red, hairless berries, up to 1cm across, which are edible but very tart.

 FLOWERING TIME March–May.

HABITAT AND DISTRIBUTION Widespread as a native plant in France, Belgium, south Germany and southern Britain in damp riverside habitats, wet woods, hedges, shaded rocks, etc., but also widely planted and naturalised throughout. Many naturalised plants are from improved horticultural varieties, but it is not always easy to be sure.

SIMILAR SPECIES Blackcurrant *R. nigrum* is very similar, but has leaves that are strongly scented when crushed (or look for the brown glands underneath), more pointed lobes, more bell-shaped flowers and black fruits. It has similar habitats and distribution, though it is often unclear where it is native and where it is naturalised.

Redcurrant flowers in spring.

Redcurrant berries.

Laburnum
Laburnum anagyroides

A small, deciduous tree, up to 7m tall, with pale, yellowish-brown, shiny bark on a slender trunk, and greyish-downy twigs. The leaves are trifoliate, up to 7cm long, greyish-green, with three narrowly ovate untoothed leaflets. The flowers are golden yellow, pea-like, individually about 2cm long but crowded into long, conspicuous, pendent racemes that may be 15–30cm long. All parts of the plant are highly poisonous and dangerous.

 FLOWERING TIME May–June.

HABITAT AND DISTRIBUTION Laburnum is native to montane regions in southern and central Europe, but widely planted in parks and gardens, and naturalised in warmer places such as northern France or southern England.

SIMILAR SPECIES Alpine Laburnum *L. alpinum* differs in being generally much less hairy, with greener twigs and longer, thinner racemes of flowers, usually 25–35cm long. It is much less commonly planted, and occasionally naturalised. The most common in gardens is hybrid Laburnum *L. x watereri*, intermediate between the other two but rarely spreading due to its sterile fruits.

Above: The striking hybrid Laburnum x watereri in flower.

Right: Alpine Laburnum flowering trees in the Alps.

Locust Tree or False-acacia
Robinia pseudoacacia

A medium-sized, erect, deciduous tree, up to about 30m high, with an open crown and, frequently, several trunks. The bark is greyish-brown, becoming deeply furrowed. It may spread by root suckers, forming dense thickets with spiny branches. The leaves are pinnately divided, up to 20cm long, with up to 10 pairs of leaflets. The flowers are pea-like, white (occasionally pink), fragrant and borne in long dense hanging racemes resembling a white Laburnum (though readily distinguished by the pinnate, not trifoliate, leaves).

 FLOWERING TIME June.

HABITAT AND DISTRIBUTION It is native to the eastern USA, but widely planted as a street and garden tree, and readily naturalising in warmer parts. The wood is notably resistant to rotting and is widely used in fencing and flooring, and the flowers are used in perfumery.

SIMILAR SPECIES *R. pseudoacacia* 'frisia' is a smaller garden form with yellowish-green foliage. Clammy Locust *R. viscosa*, also from the USA, is very sticky, with bright pink flowers.

Locust Tree flowers.

A Locust Tree in full flower.

Gorse
Ulex europaeus

A densely branched, impenetrable, very spiny shrub, up to about 2m high, often occurring in dense stands. The leaves are ternate on young plants but replaced on older bushes by long, tough, furrowed spines up to 2.5cm long. The flowers are golden-yellow, pea-like, and up to 2cm long, with long hairy sepals, and are solitary or in small clusters among the spines.

 FLOWERING TIME All year, with a peak in late winter–early spring.

HABITAT AND DISTRIBUTION Native to north-central Europe, it is often abundant or dominant on acid soil habitats such as heaths, grasslands and open woods. It is occasionally planted for hedging or ornament and spreads readily, becoming naturalised in Scandinavia.

SIMILAR SPECIES Western Gorse *U. gallii* is slightly smaller, with less deeply grooved spines, smaller flowers and a more defined flowering period in late summer. It has similar habitats but is restricted to western regions. Dwarf Gorse *U. minor* is smaller still, with shorter, softer, weakly grooved spines.

Gorse in flower.

Dwarf Gorse in flower.

Broom
Cytisus scoparius

A branched, erect, deciduous shrub, up to 2.5m tall, with slender, dark green, 5-angled twigs. The leaves are small and trifoliate lower down, simple and lanceolate above. The flowers are golden-yellow (sometimes with red patches), about 2cm long, with protruding upcurving stamens and styles. The fruit is a compressed pod, green at first and silkily hairy on the edges but hairless on the faces, which then becomes black when ripe.

 FLOWERING TIME April–June.

HABITAT AND DISTRIBUTION It is native throughout most of Europe except the far north. It is common and widespread in open and sunny habitats throughout Britain, usually on drier, more acid soils.

SIMILAR SPECIES Spanish Broom *Spartium junceum* has rounded, smooth stems, with just a few undivided leaves, which soon fall to leave rush-like erect stems. The flowers are large, up to 3cm, widely scattered along the stems and highly fragrant. A native of southern Europe, it is widely planted and occasionally naturalised in sandy, dry habitats throughout much of Europe, except the far north.

Broom in flower and fruit.

A Broom bush in full flower.

Roses

Rosa spp.

There are about 25 native species of roses *Rosa* spp. in northern Europe, together with many hybrids and a number of introduced species that may also hybridise, so it can be difficult to identify species with certainty, though it is easy enough to identify roses in general by the prickly stems, pinnate leaves and open pink or white flowers. A few groups of species, or individual species, can be identified with reasonable assurance. If the cluster of female styles in the centre of the flower is fused into a single projecting column, the species is probably either Field Rose *Rosa arvensis*, with white flowers, trailing stems and weak prickles, or Short-styled Field-rose *R. stylosa*, with pale pink flowers, on a more prickly erect bush. Plants with very downy leaves and a resinous smell are likely to be one of the downy-roses such as Sherard's Downy-rose *R. sherardii*, with deep pink flowers, which has a mainly northern distribution, or the more southern Harsh Downy-

Rosa arvensis, showing the projecting styles.

rose *R. tomentosa*. Plants that smell strongly of apple when rubbed, and have distinctly sticky leaves covered with small brown glands, are likely to be the tall, erect (up to 3m), prickly, pink-flowered Sweet-briar *R. rubiginosa*, or the smaller, paler, pink-flowered Small-flowered Sweet-briar *R. micrantha*. One particularly distinctive native rose is the Burnet Rose *R. pimpinellifolia*, a low-growing suckering shrub up to 1m tall, with erect stems covered in straight bristles and prickles. The flowers are white (occasionally pink) and the hips are black and spherical. It is widespread in dry sunny habitats throughout the area except for the far north. The most distinctive and frequently encountered non-native species is the Japanese Rose *R. rugosa*, up to 1.5m tall, covered with bristles, prickles and hairs, and with distinctly wrinkled leaves. The flowers are large, up to 7cm across, and normally bright magenta-pink. It is widely planted on roadsides and in parks and gardens, and is often naturalised.

 FLOWERING TIME May–June for most rose species.

Rosa canina hips.

Rosa pimpinellifolia in flower.

Rosa rugosa.

Rosa majalis.

Blackthorn or Sloe
Prunus spinosa

An untidy-looking, spiny shrub or small deciduous tree, up to about 6m high, usually with many ascending stems but occasionally with a single trunk. It suckers readily, and frequently forms small thickets. The bark is dark brown, often shiny, and the spreading branches end in strong, sharp spines. The leaves are ovate, dull green and up to 3cm long. The white, 5-petalled flowers appear before the leaves, solitary or paired on the branches but in such quantity as to turn the whole bush white. The fruits (known as sloes) are distinctive spherical to ovoid, blue-black, with a greyish-white bloom, up to 1.5cm across; though edible, they are very acidic.

 FLOWERING TIME March–May.

HABITAT AND DISTRIBUTION Native throughout most of Europe except the far north; common in hedges, scrub, wood margins and other sunny habitats, often forming dense stands which are very conspicuous in spring.

SIMILAR SPECIES Some forms of plum, especially Bullace *P. domestica* ssp. *institia*, are similar but generally taller, less spiny, with larger flowers that appear at the same time as the leaves, and larger fruits. It is widespread as an introduction or garden escape.

Left: Blackthorn flowers.

Right: A Blackthorn hedge in spring.

Above: Blackthorn in fruit – sloes.

Wild Cherry or Gean
Prunus avium

A medium to tall, domed, deciduous tree, up to about 30m high, with spreading branches. The bark is shiny, reddish-brown, with horizontal rows of pores, peeling horizontally in papery strips. The leaves are ovate, up to 15cm long, and on long petioles with two conspicuous red-brown glands near the blade. The flowers are white, bowl-shaped, up to 2.5cm across, in long-stalked clusters of 2–6, and appear with the leaves. The fruits are very similar to commercial cherries – flattened, spherical and ripening through bright red to dark red, though slightly smaller and less sweet.

 FLOWERING TIME April–May.

HABITAT AND DISTRIBUTION It is native throughout most of northern Europe except the far north, though widely planted for fruit and ornament almost everywhere, and often naturalised. It is common in woodlands, hedges, wood margins and other habitats on deep soils.

SIMILAR SPECIES Dwarf Cherry or Sour Cherry *P. cerasus* is similar, but generally just a shrub or small tree, with only 2–4 flowers per cluster, on shorter stalks, so the clusters are less conspicuous. It is native to south-west Asia, but widely planted for its rather acidic fruit, and often naturalised.

Left: Old Wild Cherry in autumn.

Right: Wild Cherry blossom.

Above: A Wild Cherry tree in full flower.

Bird Cherry
Prunus padus

A deciduous, rounded large bush or small tree, up to 17m high. The bark is brownish, peeling as it ages, and unpleasant-smelling. The leaves are ovate to elliptical, up to 10cm long, finely toothed along the margins and tapering to a sharp point. The flowers are white, up to 1.5cm across, 5-petalled and in long slender racemes up to 15cm long, with up to 35 flowers. The fruits are shiny bluish-black, ovoid to globose and up to 8mm long, but too sharp-tasting to be considered edible.

 FLOWERING TIME May–June.

HABITAT AND DISTRIBUTION Native to north Europe, and some mountain areas further south, it is common in hedges, scrub, wood margins and other sunny or semi-shaded habitats, tending to replace Wild Cherry in more northern areas. In Britain, it is much more common as a native in the north (and a small area of East Anglia), though it is also widely planted for ornament.

SIMILAR SPECIES Rum Cherry *P. serotina* is similar but with shinier leaves, aromatic bark and shorter, stiffer, flowering spikes. It is native to North America, but widely planted and sometimes naturalised.

Left: Bird Cherry flowering spike.

Right: Bird Cherry branch in flower.

Above: Bird Cherry bush in woodland.

St Lucie Cherry
Prunus mahaleb

A deciduous shrub or small tree up to 12m tall, though usually less. The bark is greyish-brown, with the young twigs distinctly glandular and hairy (though the older twigs soon become hairless). The leaves are ovate to almost round, up to 7cm long, with a rounded or heart-shaped base, and toothed and glandular along the margin. The flowers are white, 5-petalled, fragrant, about 1cm across, and on stalks in short racemes of up to 10 flowers. The fruit is red at first, then black, ovoid or almost spherical, up to 9mm long, fleshy and bitter.

 FLOWERING TIME April–May.

HABITAT AND DISTRIBUTION Native to south and central Europe as far north as Belgium and Germany, but widely planted elsewhere, as far north as Britain and Sweden, and often naturalised.

SIMILAR SPECIES Cherry Laurel *P. laurocerasus* is an evergreen shrub or small tree, with bright green, untoothed, leathery leaves, and white flowers in stiff racemes up to 13cm long, followed by black ovoid-spherical fruit. It is native to south-east Europe and west Asia, but very widely planted as shelter, cover and hedging, and often naturalised.

St Lucie Cherry in fruit.

St Lucie Cherry in flower.

Wild Pear
Pyrus pyraster

The 'wild pears' of northern Europe are rather complicated taxonomically, thanks to the probable combination of a native species and many ancient cultivated varieties becoming genetically intermixed. The true Wild Pear is a shrub or small tree, up to 15m high, with grey-brown hairless twigs and spiny branches. The leaves are elliptical to rounded, up to 7cm long and usually toothed towards the tip. The flowers are white, up to 3cm across, with red stamens and borne in small clusters at the same time as the leaves. The fruits are rounded to cone-shaped – in other words barely pear-shaped – yellowish-green to red in colour, up to 3.5cm long and too hard to be edible.

 FLOWERING TIME April–May.

HABITAT AND DISTRIBUTION It is native to much of Europe as far north as Belgium, France and south Germany, but widely planted and naturalised further north, including in Britain, in hedges, woodland edges and scrub.

SIMILAR SPECIES Derivatives of Pear *P. communis* can look very similar, especially if very old introductions, but they generally differ in having non-spiny branches and larger, distinctly pear-shaped fruit. It is widespread throughout the area in some form or other.

Wild Pear in flower.

Plymouth Pear
Pyrus cordata

A spiny, deciduous shrub or small tree up to 8m high, densely branched with purplish twigs. The leaves are broadly oval, up to 5cm long, finely toothed, hairy when young, then becoming dull green. The flowers are smaller than other pears, up to 18mm across, white or flushed with pink outside, in small clusters and opening at the same time as the leaves. The fruits are small, globose to cone-shaped, green, then becoming brownish-red, up to 18mm long and long-stalked when ripe. The calyx (sepals) falls off as the fruit ripens, unlike other pears where it persists even when the fruit is fully ripe.

 FLOWERING TIME April–May.

HABITAT AND DISTRIBUTION Native to western Europe as far north as south-west Britain (though British populations are now considered to probably be ancient introductions from western France) in woods, copses, hedges and scrub, often close to the sea, though never common, and occasionally planted or naturalised elsewhere in hedges and woodland edges.

Plymouth Pear in flower.

Crab Apple
Malus sylvestris

A deciduous shrub or small tree up to about 10m tall, with dark brownish-grey, scaly bark and spiny branches. The leaves are ovate to elliptical, toothed, up to 5cm long and on petioles up to 3cm long. The flowers are pinkish-white (often white inside, pink outside), fragrant, up to 4cm across and in small clusters. The fruits are typically apple-shaped, but much smaller than cultivated apples, only 2–3cm across, and hard and sour. Although not really edible, they are widely used for making jelly, preserves or cider.

 FLOWERING TIME April–May.

Crab apples in fruit

HABITAT AND DISTRIBUTION

It is native to much of northern Europe, including Britain, as far north as Finland and northern Scandinavia in hedges, copses, scrub and woodland edges; it is also occasionally planted.

SIMILAR SPECIES Forms of the cultivated apple *M. pumila* (formerly *M. domestica*) can look very similar, but are normally larger in all parts, not spiny, with larger leaves that have a relatively short petiole and larger, fleshier fruit, although there is a good deal of confusion between species, especially with ancient introductions. It is commonly planted and naturalised throughout.

Left: A Crab Apple tree in the New Forest.

Rowan or Mountain Ash
Sorbus aucuparia

A small, deciduous, slender, open tree, up to 20m high, with smooth, silvery-grey bark. The leaves are pinnate, with 5–8 pairs of oblong, toothed leaflets. The flowers are creamy-white, 5-petalled, up to 1cm across, and borne in large, domed, conspicuous clusters. The fruit is a bright, shiny, red berry, 6–9mm across, in dense clusters that are highly attractive to birds in autumn.

 FLOWERING TIME May–June.

HABITAT AND DISTRIBUTION It is native to most of northern Europe, including Britain, except for the far north. It occurs on light soils and is particularly common in upland areas, normally in sunny situations or light woodland.

SIMILAR SPECIES True Service Tree *S. domestica* is very similar, especially when in flower, but differs in having sticky buds, more fissured bark and distinctly different fruits that are pear-shaped or cone-shaped, greenish-brown or reddish, and up to 4cm long. It is native, or an ancient introduction, as far north as southern Britain, France and southern Germany, but often planted further north.

Rowan berries.

A Rowan tree in autumn.

Rowan in flower.

Whitebeam
Sorbus aria

A medium-sized tree up to 20m tall, occasionally more, with a
domed crown and smooth, grey bark. The leaves are ovate to
elliptical, up to 10cm long, with forward-pointing teeth all round the
margins and 10–14 pairs of veins, and are yellowish-green above but
conspicuously downy white below and particularly noticeable when
in bud, or in windy weather when the undersides are exposed.
The flowers are white, 10–15mm across, in dense, rather flat heads,
followed by bright red, ovoid fruits, 10–15mm long.

 FLOWERING TIME May–June.

Whitebeam leaves unfurling.

Right: Whitebeam tree in flower,
on limestone.

HABITAT AND DISTRIBUTION It is native to much of northern Europe, including Britain, though its exact distribution is unclear, as it is now known to consist of a complex collection of micro-species, each often confined to a small area. It is also widely planted and occasionally naturalised.

SIMILAR SPECIES If all the microspecies are considered, there are 30 or more similar species needing detailed examination for certain identification. Swedish Whitebeam *S. intermedia* has more deeply lobed leaves, cut about one-third of the way to the midrib, and more yellowish-white under the leaves. It is native to north Germany and Scandinavia, but widely planted and naturalised throughout.

Wild Service Tree
Sorbus torminalis

A medium-sized, spreading, domed, deciduous tree, up to 25m high, with greyish, fissured bark. The leaves are roughly triangular to ovate in outline, up to 10cm long, but deeply lobed into 3–5 pairs of triangular, pointed lobes, looking much more like a maple leaf than a normal *Sorbus*. The leaves are green on both surfaces but become orange-brown to bright red in autumn before falling. The flowers are white, 10–15mm across, on woolly stalks, in domed, open clusters, followed by the ovoid brownish-green dotted fruits up to 18mm long. The fruits are edible, and said to resemble dates, but need to be over-ripe to remove their acidity.

 FLOWERING TIME May–June.

HABITAT AND DISTRIBUTION It is native throughout most of Europe as far north as Denmark, in woods, scrub and hedgerows, most often on heavier calcareous soils. It is native in Britain as far north as the Scottish border, but planted elsewhere.

SIMILAR SPECIES It is distinct from all other *Sorbus* species, but the leaf shape could be confused with maples, though they have quite different flowers and fruits (see pp. 142–146).

Right: Wild Service Tree in flower.

The distinctive leaves of Wild Service Tree.

Hawthorn or Quickthorn
Crataegus monogyna

A dense, deciduous shrub or small tree, either bushy from the base or with a distinct trunk, with brownish-grey, fissured, peeling bark. The densely branched twigs are liberally covered with strong, sharp spines, up to 2.5cm long. The leaves are dark, shiny green above, hairy below, ovate with a wedge-shaped base and deeply lobed to more than halfway to the midrib into 5–7 lobes that are themselves toothed or lobed towards their tips. The flowers are white or pink, about 1.5cm across, with a single, central style, and are gathered into clusters of 10–18 flowers, which are strongly scented. The fruits are globose to ovoid, bright red, up to 1.5cm long, with one 'pip', and often produced in great abundance in autumn. It is also known as May or May Blossom.

 **FLOWERING TIME** May–June.

HABITAT AND DISTRIBUTION It is native throughout most of Europe, except for the far north, in woods, scrub and hedgerows on a variety of different soils. It is also widely planted for hedges and shelter.

SIMILAR SPECIES *C. calycina* is a native of mainland Europe (not Britain), north as far as south Scandinavia, which differs in its conspicuous, erect calyx persisting on the fruits. See also Midland Hawthorn, next page.

Hawthorn blossom.

Hawthorn bush in full flower.

Hawthorn berries, or haws.

Midland Hawthorn
Crataegus laevigata

A deciduous shrub or small tree, up to 10m tall. It is generally similar in appearance to Common Hawthorn (see p. 72), but differs in being less spiny, with leaves much less deeply divided, and normally just lobed towards the tip; the leaves are more or less hairless below. The flowers have 2–3 styles and the fruit has 2–3 nutlets (always one style and one nutlet in Common Hawthorn).

 FLOWERING TIME May–June.

HABITAT AND DISTRIBUTION It is native throughout most of Europe as far north as Denmark and Norway in woods, scrub and hedgerows, and is more tolerant of shade than Hawthorn. It is native in southern Britain, but it is planted elsewhere.

Midland Hawthorn in flower.

Shrubby Cinquefoil
Potentilla fruticosa

A small, deciduous, erect or spreading shrub up to 1.5m high, with slender, twisted branches covered by the remains of old leaf bases. The leaves are small, silvery greyish, and pinnate with 5–7 narrowly lanceolate, untoothed leaflets. The flowers are large, 2–3cm across, bright yellow, with the petals rounded and much longer than the sepals, and are solitary or in small clusters.

 FLOWERING TIME May–July.

HABITAT AND DISTRIBUTION
It is native in several separate parts of northern Europe, mainly northern England, western Ireland, and parts of Norway and Sweden in sunny, damp or dry, calcareous habitats, including limestone pavement.

Right: Shrubby Cinquefoil in western Ireland.

Sea-buckthorn
Hippophae rhamnoides

A spiny, deciduous, much-branched, dioecious shrub, readily producing suckers, normally up to about 3m, though rarely forming a small tree. Young twigs are covered in silvery scales, later becoming brown. The leaves are narrowly linear, up to 8cm long and about 1cm wide, and covered with silvery or pale brown scales when young. Male flowers are inconspicuous, with 4 stamens, produced before the leaves. Female flowers (on separate plants) are small, greenish-yellow, and in axillary clusters before the leaves. The berries are much more visible, bright orange, globose, up to 8mm across and often produced in abundance. They are edible, though rather tart, and considered to be a delicacy in some areas.

 **FLOWERING TIME** March–May.

HABITAT AND DISTRIBUTION It is native throughout most of Europe, including Britain, as far north as Scandinavia. It is most common in coastal and montane habitats, though also occurring in inland grasslands, and widely planted as a stabiliser on dunes or along roadsides, frequently naturalising.

The striking orange berries of the Sea-buckthorn in autumn.

Sea-buckthorn in fruit.

Buckthorn or Purging Buckthorn
Rhamnus cathartica

A deciduous, much-branched, spiny, dioecious shrub, up to 10m high, though normally less. The bark of older trees is black, peeling to reveal paler patches. The leaves are ovate to elliptical, up to 7cm long, and finely toothed around the margin, with a short petiole. The flowers are greenish, 4-petalled (occasionally 5), small, and in tight clusters in the axils. Male and female flowers are usually on separate plants, but not necessarily. The berries are globose, up to 8mm across, and ripening to shiny black, in little clusters. It is one of the few larval food plants of the Brimstone butterfly, which lays its eggs on the developing leaves.

 FLOWERING TIME May–June.

HABITAT AND DISTRIBUTION It is native throughout north Europe as far north as southern Sweden in hedges, scrub, grasslands and wood margins, almost always on calcareous soil. It is native to England and Wales, but planted elsewhere.

SIMILAR SPECIES Mediterranean Buckthorn *R. alaternus* is similar but not spiny, with flower parts in fives. It is native to southern Europe, but occasionally planted and naturalised in places.

Ripe Buckthorn berries.

Buckthorn in flower.

Alder Buckthorn
Frangula alnus

An erect, unspiny, deciduous, bushy shrub or small tree, up to about 5m high (rarely, 10m). The bark on older parts is greyish-brown, becoming fissured and marked with purple, while younger twigs are green at first, becoming grey or brown with prominent leaf scars and pointed dark buds. The leaves are roughly oval, broadest above the middle and shiny green above, and thin, with untoothed margins; in autumn, they turn bright yellow or reddish. The flowers are small, inconspicuous, about 3mm across, solitary or clustered on younger wood, 5-parted, with both male and female parts together (i.e. hermaphrodite). The berries are globose, up to 1cm across, yellowish at first, then red, and finally black, often with all stages together. Alder Buckthorn is the other main larval food plant of the Brimstone butterfly (see p. 78).

 FLOWERING TIME May–June.

HABITAT AND DISTRIBUTION It is native throughout northern Europe as far north as southern Scandinavia in woodland, bogs and scrub, normally on damp, peaty, slightly acidic soils. It is native in England and Wales, but rare or planted elsewhere.

Alder Buckthorn in flower.

Berries and autumn colour.

The Elms
Ulmus spp.

The elms are readily recognised as a group, as trees with large, oval leaves that have distinctly asymmetrical bases, inconspicuous clusters of small, normally hermaprodite flowers produced before the leaves, and circular or oval winged nuts, notched at the tip. However, individual species can be very hard to identify thanks to the existence of numerous hybrids, intermediates and local populations, giving rise to a good degree of disagreement amongst taxonomists. Some of the most distinctive in our area include the Wych Elm *Ulmus glabra*, a broad tree up to 40m, normally without suckers. The leaves are large and up to 16cm long, with 12–18 pairs of lateral veins. It is native throughout most of north Europe except the far north; within Britain, it is commonest in the north and west. English Elm *U. procera* is a suckering tree to 30m (though very few mature trees now exist due to the effects of Dutch Elm disease), often with corky bark; the leaves are smaller, rounded, up to 9cm long, and rough above, with 10–12 pairs of veins. It occurs throughout Britain in hedgerows, either native or planted, and is rare elsewhere in Europe. Small-leaved Elm *U. minor* is made up of several related

Wych Elm leaves.

Wych Elm in fruit.

species, characterised by suckering trees, ovate leaves that are smooth above, with 7–12 pairs of lateral veins, and oval, winged fruits. It is native throughout northern Europe except the far north; within Britain, it is largely confined to eastern England. A number of non-native species are also planted within the area.

English Elm in fruit, showing the corky twigs.

Common Mulberry
Morus nigra

A small tree up to about 14m high, with a gnarled and twisted trunk and intricate branching, giving the impression of an old tree even when it's not old. The leaves are broadly oval with a heart-shaped base, up to 20cm long, and variable between simple and deeply lobed. The flowers are greenish-yellow in short spikes, with males and females separate. The fruits are rather like raspberries, ripening to dark red, and are often produced in great abundance.

 FLOWERING TIME May.

HABITAT AND DISTRIBUTION Native to Asia, but long planted in many parts of Europe for its fruit, which is used mainly for preserves.

SIMILAR SPECIES White Mulberry *M. alba* is similar, with much paler fruit, and it was formerly widely planted as the food for silkworms.

Common Mulberry leaf and fruit.

Beech
Fagus sylvatica

A large, elegant, broad-domed, deciduous tree, up to about 40m high, with a well-defined trunk covered with smooth, grey-brown bark. The buds are distinctive reddish-brown, up to 2cm long, narrow, cylindrical and pointed. The leaves are ovate to elliptical, barely toothed, and light green with a silky margin at first, but gradually becoming darker and more leathery. The male flowers are in long-stalked, hanging, greenish-yellow clusters, while the female flowers (on the same tree) are inconspicuous, paired on short stalks. The fruits are the familiar 'beech mast', 1–3 shiny, deep brown, 3-sided nuts enclosed in a woody, spiny husk.

FLOWERING TIME April–May.

A planted Beech avenue in winter.

Beech flowers.

Autumn Beech leaves.

Beech seedling.

HABITAT AND DISTRIBUTION It is native throughout most of Europe (though confined to mountains further south) as far north as southern Norway, most commonly, though not necessarily, on calcareous soils. It is native in Britain in southern England and Wales, but very widely planted elsewhere, and often naturalised.

SIMILAR SPECIES Copper Beech *F. sylvatica* 'purpurea' is similar, but has dense, dark purple foliage, and is widely planted in parks and gardens.

An ancient Beech tree in spring.

Sweet Chestnut
Castanea sativa

An attractive, large, spreading, domed, deciduous tree, up to 30m tall. In younger trees, the bark is silvery and smooth, gradually developing vertical fissures, but mature trees have a substantial trunk with noticeably spiralling vertical grooves or fissures. The leaves are large, up to 25cm long, narrowly oblong, glossy above, with regularly sharp-toothed margins. The flowers consist of a long, spreading to pendulous inflorescence. The male flowers are clustered into a long, yellowish-white terminal catkin with little visible except stamens, and with a few small clusters of greenish female flowers at the base of the catkin. These later develop into the familiar sweet chestnuts – 3 or so large, rich brown, shiny nuts enclosed within a bright green, highly spiny, splitting capsule.

 FLOWERING TIME June–July.

HABITAT AND DISTRIBUTION A native of an uncertain area of southern Europe, but it is very widely planted except in the far north for timber, nuts or ornament, and frequently naturalised in warmer areas. It is not native in Britain, but planted almost throughout; in southern England it naturalises readily and may form woodlands.

Left: Sweet Chestnut flowers.

Right: Chestnut woodland in autumn.

Ripe Sweet Chestnut fruits.

Pedunculate Oak, English Oak or Common Oak
Quercus robur

An attractive and familiar, large, deciduous tree up to 40m high, with a heavy, domed crown. The bark of a mature tree is greyish-brown and fissured, or browner and smoother when young. The leaves are roughly oval, up to 12cm long, with several pairs of wavy, rounded lobes, a very short stalk (less than 5mm long), and distinct ear-like auricles at the base. The male flowers are produced in clusters of long, thin, pendulous catkins, composed mainly of stamens, up to 10cm long. The female flowers are greenish, inconspicuous, unstalked, and in groups of 1–3. The fruit is the familiar acorn – an ovoid, woody, greenish nut borne in a scaly cup that is up to 1.8cm wide, on a stalk that is at least 4cm long.

 FLOWERING TIME April–May.

Above: A very ancient English Oak in Dorset.

Right: An old hedgerow English Oak.

English Oak acorns.

English Oak male catkins.

English Oak leaf.

HABITAT AND DISTRIBUTION It is native throughout northern Europe (and elsewhere) except for the far north, forming woodlands in which it is often dominant, most commonly on heavy soils. It is widely planted almost everywhere, and the boundaries between native and introduced have become blurred.

SIMILAR SPECIES White Oak or Downy Oak *Q. pubescens* has smaller leaves, and the foliage and twigs are very downy. It is mainly southern, native as far north as south Germany, but planted elsewhere including the UK.

Sessile Oak or Durmast Oak
Quercus petraea

A large, deciduous tree, up to about 40m high, with a domed crown. It is rather similar in general appearance to Pedunculate Oak, but with a generally longer, straighter trunk and straighter branches. The bark is greyish and fissured. The leaves are roughly oval and similar in general size and shape to those of Pedunculate Oak, but the blade narrows into a wedge-shaped base without auricles, and the petiole (stalk) is at least 1cm long, or up to 2.5cm. They are hairless above, but distinctly hairy on the underside. The flowers and fruit are generally similar to those of Pedunculate Oak, but the acorns are more or less stalkless (the Sessile of the name refers to the acorns, not to the leaves).

FLOWERING TIME April–May.

An old Sessile Oak tree.

Turkey Oak tree.

HABITAT AND DISTRIBUTION It is native almost throughout Europe except for the far north and parts of the south. It readily forms and dominates woodlands, especially on lighter, more acidic soils, and in the uplands. It is native throughout Britain, but much more common in the west.

Sessile Oak leaf.

Sessile Oak acorns.

Sessile Oak male catkins.

Fallen acorns and leaves of Turkey Oak.

SIMILAR SPECIES Turkey Oak *Q. cerris* is similar, but has leaves with more pointed lobes. The buds are surrounded by long, brownish stipules, and the acorn cup has reflexed scales. It is native to southern Europe, but widely planted.

Holm Oak or Evergreen Oak
Quercus ilex

A large, dense, evergreen, broadly domed tree, up to about 25m high, with a short trunk covered in dark grey bark that cracks into roughly square plates. The leaves are evergreen, up to about 8cm long, variously lobed or spiny, with a wedge-shaped base, a dark green upper surface and a paler downy underside. The leaves on young trees or branches are more likely to be spiny and shaped like holly – hence the scientific name *ilex* and the English name Holm. The male catkins are yellowish-green, borne in pendulous clusters in spring; the acorns are small and in scaly cups that enclose about half of the acorn.

 FLOWERING TIME April–May.

HABITAT AND DISTRIBUTION It is native to southern Europe but widely planted in the warmer parts of northern Europe, especially near the coast, where it may naturalise in favourable conditions.

Western form of Holm Oak.

A planted Holm Oak tree.

Holm Oak leaves and acorns.

Cork Oak tree.

SIMILAR SPECIES Cork Oak *Q. suber* is rather similar but differs in being a smaller and more open tree, up to 20m high, with a conspicuously thickened, corky bark (the 'cork' of commerce). The leaves and flowers are similar, but the acorn cups have distinctly reflexed projecting scales. It is native to south-west Europe, but widely planted in the warmer parts of northern Europe for cork, ornament or curiosity.

Walnut
Juglans regia

An attractive, medium-sized, deciduous tree with an open-domed crown, usually arising from a straight trunk. The bark is greyish and smooth at first but becomes fissured on older trees. The leaves, up to 15cm long, are pinnately divided into 5–9 ovate, untoothed, leathery, green leaflets in several pairs with a large terminal leaflet. They are reddish when they first appear in spring and become yellow in autumn. The male catkins are yellowish-green, pendulous and up to 15cm long, while the female flowers are small, greenish, in little clusters, and covered with greenish hairs. The fruit is ovoid, shiny, green, fleshy, up to 5cm long and contains the familiar, edible, wrinkled walnut.

 FLOWERING TIME April–May.

HABITAT AND DISTRIBUTION It is native to south-east Europe and Asia, but widely planted throughout the warmer parts of Europe, in plantations further south, or more casually in the north. It is naturalised except in the far north.

SIMILAR SPECIES Black Walnut *J. nigra* from North America is similar, but the leaves are divided into more and smaller leaflets – up to 10 pairs with a small terminal leaflet. The fruits are edible but are rarely used. It is planted in places from Denmark southwards and rarely naturalised.

Left: Walnut fruits.

Right: A walnut tree in autumn.

Bog-myrtle
Myrica gale

A small, erect shrub, up to 2m high, usually much-branched, with reddish-brown twigs that have yellow glands. The leaves are roughly elliptical, toothed towards the tip, up to 6cm long, glandular and strongly aromatic. The catkins are reddish-brown, up to 4cm long and appear in early spring on the previous year's naked branches before the leaves open. The fruit is a narrowly 2-winged nut. The bushes may have just male flowers, just female flowers, or both, and may even change during their life. Bog-myrtle was once used to flavour a beer, and its strongly aromatic foliage is now being used for the manufacture of an insect repellent.

 FLOWERING TIME March–May.

HABITAT AND DISTRIBUTION Native throughout northern Europe, except the far north, in damp, generally acidic habitats such as bogs, wet heathland and moorland.

SIMILAR SPECIES
Bayberry *M. pensylvanica* is larger, with hairier young twigs, catkins borne on the new growth, and fruit a small, waxy berry. It is native to the eastern USA, sometimes planted for cover and occasionally naturalised.

Left: Bog-myrtle male catkins.

Right: Bog-myrtle in fruit.

Silver Birch
Betula pendula

A medium-sized, slender, graceful, open tree, with a narrow, rounded crown and noticeably down-swept branches on older trees. The narrow trunk is covered with silvery white bark, which becomes fissured, darker and separated into plates at the base in older trees, while the young branches are silvery brown. The leaves are ovate-triangular, up to 7cm long, hairless and with doubly serrate margins; that is, there are a number of larger triangular teeth, each of which is itself toothed. The male catkins are yellowish, narrowly cylindrical, pendulous, and in small clusters at the end of twigs. They first appear in winter, when they are greenish, then mature as the leaves are opening. The female catkins (on the same tree) are shorter, green, erect and grow from the leaf axils, lengthening after flowering to produce large quantities of papery, brown, winged seeds. The trees are shallow-rooted and relatively short-lived.

 FLOWERING TIME April–May.

HABITAT AND DISTRIBUTION Native almost throughout northern Europe, except for the far north, on dry acidic sites such as heathlands and occasionally forming woods. It is a pioneer species that is often replaced by other trees as the woodland matures. It is common throughout most of Britain, both as a native and planted.

SIMILAR SPECIES See Downy Birch, p. 104.

Left: Silver Birch leaf.

Right: Silver Birch tree in autumn.

Downy Birch
Betula pubescens

A similar tree to Silver Birch, it is up to about 25m high, though rather more untidy. The bark is brown or greyish – rarely bright silvery white – without the fissuring and plates of Silver Birch. The branches are more erect, less pendulous at the tips and are downy when young. The leaves are more rounded, not long-pointed, with evenly distributed teeth on the margins. The flowers and fruit are similar, though the seeds have smaller wings.

 FLOWERING TIME April–May.

HABITAT AND DISTRIBUTION Native almost throughout northern Europe except for a few lowland areas, it favours damper, poorly drained, acid soils (where it normally replaces Silver Birch); it is rarely planted.

SIMILAR SPECIES Paper-bark Birch *B. papyrifera* usually has shining, white, peeling bark, and large leaves and catkins. It is from North America and often planted. Himalayan Birch *B. utilis* has variously coloured bark and large leaves that are shiny above. It is from the Himalayas and planted for ornament in most countries of north Europe.

Left: Downy Birch trunks and bark.

Right: Downy Birch leaf.

Old Downy Birch tree.

Dwarf Birch
Betula nana

A dwarf shrub, rarely exceeding 1m high – though a genuine shrub – with spreading or ascending branches. The leaves are almost orb-shaped, rounded at both ends and up to 2cm across, with regular rounded teeth. The catkins are small, erect and 1–2cm long in fruit.

 FLOWERING TIME June–July.

HABITAT AND DISTRIBUTION It is native in northern Britain (from Yorkshire northwards), Scandinavia and parts of the Alps, in tundra, moorland and edges of bogs. It is locally common where it does occur and rarely planted.

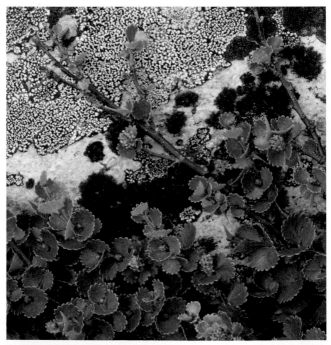

Dwarf Birch in flower and fruit.

A Dwarf Birch in fruit.

Green Alder
Alnus viridis

A small shrub, up to 5m high, though usually less. The leaves are bright green, oval to elliptic, pointed, sticky, finely double-toothed on the margin and hairy on the veins below. The catkins appear 2 or 3 together on short shoots, with a few leaves just below them. Male catkins are pendulous, while females are erect and narrowly ovoid.

 FLOWERING TIME April–June, or later at higher altitudes.

HABITAT AND DISTRIBUTION It is native in a variety of damp mountain habitats in the Alps and Jura, planted elsewhere for ornamental reasons, and occasionally naturalised.

SIMILAR SPECIES Italian Alder *A. cordata* is a tree up to 25m high, and has leaves with heart-shaped bases and long points. Male catkins are very long. It is widely planted and occasionally naturalised.

Left: Green Alder branch in spring.

Right: Italian Alder male and female cones.

Alder
Alnus glutinosa

A small to medium-sized deciduous tree, commonly up to 20m high, occasionally up to 30m, with a broad, domed crown. The bark is dark brown, becoming fissured and separated into plates as it matures. The buds are brown and sticky when young (hence the scientific name *glutinosa*). The leaves are broadly ovate to almost round, wedge-shaped or heart-shaped at the base and rounded or notched at the tip, and up to 10cm long. The male flowers first appear in winter as clusters of 2–3 purplish pendulous catkins, which open to reveal yellow pollen in spring. The female catkins are small, in little stalked clusters that are purplish-red, becoming green. These mature into small, woody, cone-like fruits, up to 3cm long, that disperse quantities of small, winged seeds providing vital winter food for various finches such as siskins.

 FLOWERING TIME February–April.

HABITAT AND DISTRIBUTION It is native almost throughout northern Europe except for the far north, in damp places and especially along rivers and around marshes.

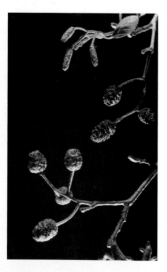

SIMILAR SPECIES Grey Alder *A. incana* is similar, but the leaves are pointed and not sticky when young, and paler below than above. It is native over much of northern Europe (but not Britain), and also planted occasionally for timber.

Left: Female Alder cones.

Right: Alder tree in early winter.

Hornbeam
Carpinus betulus

A medium-sized, deciduous tree, up to 25m tall, with a rounded outline. The trunk is often twisted and gnarled, with silvery grey bark, and frequently fissured towards the base. The leaves are up to 10cm long, oblong to ovate, pointed, sharply toothed, with a distinctly pleated appearance. In autumn they turn an attractive yellowish colour. The male flowers are in catkins, up to 5cm long, yellowish-green with red-edged scales, and appear with or just before the leaves. The fruit are distinctive and conspicuous, in a pendulous cluster up to 15cm long, with up to 12 pairs of nuts, each with long, 3-lobed bracts, of which the middle lobe is much the longest. The seeds are much-favoured by hawfinches.

 FLOWERING TIME April–May.

HABITAT AND DISTRIBUTION It is native throughout much of northern Europe as far north as southern Sweden in woodlands, hedges and common land, and is widely planted for timber. It is native to southern England and Wales and introduced elsewhere in Britain.

SIMILAR SPECIES Hop Hornbeam *Ostrya carpinifolia* has similar foliage and structure, but the fruiting clusters are tighter, with untoothed bracts resembling hops. It is native to southern Europe, but often planted.

Left: The fruit and bracts of Hornbeam.

Right: Hornbeam wood in Kent.

Spindle
Euonymus europaeus

A much-branched, hairless, deciduous shrub or very small tree, up to 6m high, occasionally more. The bark is smooth and greyish, becoming more fissured with age. The twigs are green, angled when young and becoming rounder with age. The leaves are elliptic to ovate, dark green, finely toothed or untoothed, pointed, up to 10cm long and turn yellow or bright red in autumn. The flowers are greenish-white, usually 4-parted, in small clusters in the leaf axils. The fruits are highly conspicuous, consisting of a bright magenta-pink 4-angled outer husk enclosing 4 bright orange berries that become exposed as the fruit matures.

 FLOWERING TIME May–June.

HABITAT AND DISTRIBUTION It is native almost throughout northern Europe, except the far north, in woods, hedgerows and scrub, particularly on calcareous soils such as limestone.

Spindle bush in autumn.

Spindle flowers.

SIMILAR SPECIES Broad-leaved Spindle *E. latifolia* is very similar, but has more obviously toothed leaves, pinkish flowers with 5 petals, and sharply angled, 5-chambered fruits. It is native to south and central Europe, but planted for ornament elsewhere. Evergreen Spindle *E. japonica*, with leathery evergreen leaves, is widely planted for hedging, especially in coastal regions.

Spindle fruit splitting to reveal the berries.

Hazel

Corylus avellana

Most commonly a spreading, multi-stemmed shrub up to about 6m high, though occasionally forming a small tree up to 12m. The bark is pale brown, smooth, shiny and peeling in strips, and the young twigs are hairy. The leaves are almost round in shape, up to 10cm long, pointed at the tip and heart-shaped at the base, with a double-toothed margin. Male catkins first appear in early winter, then open in spring to become the familiar pendulous, yellow, 8cm-long, pollen-rich catkins, opening well before the leaves. The female flowers are distinctive but inconspicuous – they consist of a large bud with a starry circle of small protruding red styles, resembling a tiny sea anemone. These develop into the familiar hazelnut – a hard-shelled, ovoid nut in bunches of 1–4 surrounded by a husk of fused, leafy, raggedly toothed bracts. These are popular food for a variety of mammals – each leaving distinctive marks – from dormice to squirrels.

 FLOWERING TIME February–April.

HABITAT AND DISTRIBUTION It is native almost throughout Europe, except for the far south and the far north, in woods, hedges and scrub on a variety of soils. It is frequently managed as coppice, by regular cycles of cutting, to produce wood for a variety of uses.

Left: Hazel leaves and young catkins in autumn.

Right: Fallen hazelnuts.

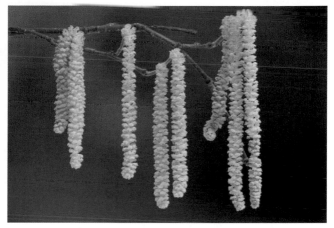

Hazel catkins in spring.

SIMILAR SPECIES Filbert *C. maxima* is generally similar, except that the nuts are completely enclosed by a longer, tubular involucre that constricts beyond the nut. It is native to south-east Europe and is often planted for its nuts, which are more widely used than those of Hazel, including those known as Kentish Cobs. Turkish Hazel *C. colurna* is much larger, sometimes becoming a tree up to 25m tall, and the nuts are enclosed by long, deeply toothed, raggedly cut, reflexed bracts. It is native to south-east Europe and south-west Asia, but widely planted for ornament and for its nuts.

White Poplar
Populus alba

A strong-growing, erect, suckering, deciduous tree, with a broad crown. The bark is smooth and silvery white when young and becomes darker and rougher with age, with white-downy twigs. The leaves are distinctive, stalked, triangular to ovate in shape, deeply palmately lobed, and dark green above but bright downy-white below. In windy weather, all the white undersides are conspicuously revealed. The catkins are pendulous, up to 8cm long, or longer in fruit. The male catkins are rarely seen. Female catkins produce large quantities of light, fluffy seeds.

 FLOWERING TIME March–April.

HABITAT AND DISTRIBUTION It is native in central and eastern Europe in damp places, especially river valleys, and widely planted, especially near the coast.

Above: White Poplar leaves showing the pale underside.

Right: White Poplar tree in spring.

Black Poplar
Populus nigra

A large, heavy, spreading, deciduous tree, up to about 30m high, with heavy arching branches and a thick trunk covered with bosses or burrs. The bark is light greyish-brown, uneven and fissured when old. The leaves are roughly oval- to diamond-shaped, finely toothed and long-pointed, with a flattened stalk. The male flowers are in slender pendulous catkins, about 5cm long, greyish at first but becoming scarlet as the abundant stamens open up. The female catkins are greenish and up to 15cm long when in fruit.

FLOWERING TIME February–April.

HABITAT AND DISTRIBUTION It is native from Britain, Holland and Germany southwards in damp situations, especially river floodplains and wet woodland, but widely planted and naturalised elsewhere. It is native but uncommon in England and Wales and planted elsewhere in the British Isles.

SIMILAR SPECIES The native subspecies is ssp. *betulifolia*. Other forms occur as planted trees, most notably the Lombardy poplar *P. nigra* 'italica', which is tall and columnar in shape, with no burrs on the trunk, and widely used as an ornamental tree, especially in avenues.

Left: The trunk and bark of Black Poplar.

Right: An old floodplain Black Poplar tree in Dorset.

Aspen
Populus tremula

A slender, domed or conical deciduous tree, up to about 25m high. The trunk is quite slender, often leaning slightly, with smooth grey-green bark that becomes rougher with age, suckering freely around the base to form dense thickets of young trees. The leaves are broadly ovate to almost circular, rounded at the tip, wavy-edged with blunt teeth and a long flattened petiole up to 6cm. Aspen is one of the few trees that you can identify by sound, as the leaves flutter noisily and continuously in the slightest breeze! The male catkins are pendulous, with purplish-red stamens amongst silvery bracts. The female catkins (on separate trees) are greenish, up to 12cm long and produce fluffy seeds that germinate very rapidly.

FLOWERING TIME February–April.

HABITAT AND DISTRIBUTION It is native almost throughout Europe, except for the far north, in a variety of situations, especially where the soil is damp, and in mountain areas. It is a short-lived, colonising tree, rarely dominating a woodland.

SIMILAR SPECIES American Aspen *P. tremuloides* is very similar, but the leaves are more pointed and finely toothed. It is native to North America and occasionally planted.

Left: Male Aspen catkins in spring.

Right: Aspen trees in autumn.

Crack Willow
Salix fragilis

A medium to large deciduous tree with a broad, tapering crown. The trunk is covered with blackish-grey, scaly bark, becoming criss-crossed with ridges and fissures, and arching branches that arise low down on the trunk. The young shoots are olive-green and very fragile, snapping off readily at the joints. The leaves are narrowly lanceolate, up to 15cm long and 2–4cm wide, coarsely toothed on the margins, glossy green above, and paler and slightly hairy below. The catkins appear at the same time as the leaves. The male catkins are yellow, cylindrical, up to 5cm long and drooping. The female catkins, on separate trees as with all the willows, are smaller and green, and later produce the fluffy windborne seeds. The broken-off twigs root readily and are one means of distribution for the tree, especially for riverside trees.

A Crack Willow tree in autumn.

Crack Willow male catkins.

FLOWERING TIME April–May.

HABITAT AND DISTRIBUTION It is native throughout northern Europe, except the far north, in damp riverside or lake margin habitats and is considered to be an ancient introduction in some parts of its range.

SIMILAR SPECIES A number of hybrids and varieties exist. It is most easily confused with White Willow (see p. 126 for differences).

White Willow
Salix alba

A rather similar tree to Crack Willow, but with less fragile twigs that do not readily break off. The leaves are smaller, up to 10cm long and 1.5cm wide, and green above (hairy at first) but distinctly persistently silvery white and hairy below. The catkins are broadly similar to those of Crack Willow. It is often pollarded, producing short, thick trunks topped with a mass of young branches.

 FLOWERING TIME April–May.

HABITAT AND DISTRIBUTION It is native throughout northern Europe, except the far north, in damp meadows, marshes and riversides. It is also widely planted and the natural distribution has become blurred.

Above: Leaves of White Willow.

Right: White Willow tree in late summer.

Goat Willow or Great Sallow
Salix capraea

The small, shrubby willows, with the early spring 'pussy willow' flowers, are a complex and difficult group, with many hybrids and intermediates, from which a few key species are described. Goat Willow is a shrub or small tree, normally under 10m high, but occasionally up to 20m, with coarsely fissured, brown bark. The leaves are oblong to oval, about twice as long as broad, with a short, pointed tip and white-hairy underneath. The stipules are small, heart-shaped and usually fall early. The male catkins are roughly ovoid, up to 3cm long, with a mass of yellow stamens emerging from silky grey hairs that are produced before the leaves open, and are often known as 'pussy willows' – a familiar harbinger of spring. The female catkins, on different plants, are greener and smaller.

 **FLOWERING TIME** March–April.

HABITAT AND DISTRIBUTION It is native throughout northern Europe, except the far north, in damp habitats including carr woodland, marshes and wet moorland. It is locally common and it is the key larval food plant of the Purple Emperor butterfly.

An ancient Goat Willow tree on Dartmoor.

Above: Goat Willow male flowers – 'pussy willows'.

Left: The leaves and stipules of Eared Willow.

SIMILAR SPECIES Grey Willow or Sallow *S. cinerea* has smaller leaves, usually 3–4 times as long as they are broad, and hairy below, often with rusty-red hairs. The 2-year-old twigs have a series of fine longitudinal ridges under the bark (best observed by peeling off the bark), which are absent from Goat Willow. The flowers are very similar to those of Goat Willow, though usually slightly smaller. It is abundant throughout the area in a variety of damp habitats. Eared Willow *S. aurita* is a shrub that grows up to 3m, with broadly ovate leaves that are conspicuously wrinkled or rugose above and have wavy margins and large persistent ear-like stipules. It is moderately common in wet acidic places throughout the area, except the far north.

Osier
Salix viminalis

A shrub or small tree up to 5m tall (occasionally 10m where it is allowed to grow to its full height, though often it is cut back to produce young flexible shoots), with markedly flexible, yellowish-brown twigs. The leaves are long and narrow, up to 15cm long by 1.5cm wide, green above but silky-hairy below, with untoothed turned-over margins. The catkins are narrow, up to 3cm long, and crowded towards the tips of shoots, appearing just before the leaves.

 **FLOWERING TIME** March–April.

HABITAT AND DISTRIBUTION It is native in central Europe as far north as Britain, Belgium, Holland and Germany in damp habitats including carr woodland, marshes and riversides. It is very widely planted as the key species for basket-making and weaving, and often naturalised.

SIMILAR SPECIES Violet Willow *S. daphnoides* is a shrub or small tree growing up to 10m. The branches are shiny brown but have a bluish, waxy bloom when young. The leaves are oblong-lanceolate, up to 12cm long, finely toothed and without turned-over margins. It is native in a variety of damp habitats in Scandinavia, but it is often planted elsewhere for ornament and occasionally naturalised.

Above: Osier leaves.

Left: Osier bush in autumn.

Downy Willow
Salix lapponum

A low-growing, much-branched shrub up to 1.5m tall, with twigs that are grey-downy at first and later become hairless and dark reddish-brown. The leaves are oval to lanceolate, up to 7cm long and 2.5cm wide, somewhat hairy above but densely hairy underneath, similarly coloured above and below, and untoothed or slightly toothed. The catkins are rather rigid, up to 4cm long, and appear with or just before the leaves, with long, white hairs amongst the flowers.

 FLOWERING TIME May–June.

HABITAT AND DISTRIBUTION Found in northern Britain and throughout Scandinavia in tundra, on rock ledges and on damp mountain slopes up to 1,100m high. It is uncommon in northern Britain. It is also cultivated in gardens.

Downy Willow bush in the Cairngorms.

Downy Willow leaves and fruit.

SIMILAR SPECIES Woolly Willow *S. lanata* has broader leaves up to 7cm long but almost as broad at up to 6.5cm wide, that are densely white-downy at first but soon become hairless. It appears in similar mountain and Arctic habitats up to 1,700m throughout Scandinavia, mainly in base-rich sites. It is very uncommon in Scotland in the central Highlands. It is also grown in gardens and is occasionally naturalised.

Weeping Willow
Salix x sepulcralis

A distinctive, elegant, broad tree up to 20m high, with long, pendent, yellowish or greenish shoots that droop almost to ground level. The leaves are narrowly lanceolate, up to 12cm long, pointed, finely toothed and hairy when young but soon become hairless. The catkins are slender, yellow, up to 3cm long and appear with the leaves.

 FLOWERING TIME April.

HABITAT AND DISTRIBUTION A familiar tree, of garden origin, originally from China, that is widely planted for ornament and occasionally naturalised.

SIMILAR SPECIES Weeping Crack-willow *S. x pendulina* has a similar form, but with broader, hairless leaves and more fragile shoots. It is occasionally planted and sometimes naturalised.

Weeping Willow tree (with Golden Willows in foreground).

Net-leaved Willow or Net-veined Willow
Salix reticulata

A dwarf shrub, barely recognisable as a shrub, that grows up to about 25cm high. The leaves are distinctive, oval to almost round, up to 4cm long, untoothed, shiny, dark green and with a conspicuous network of veins. The catkins are slender, cylindrical, up to 3cm long, purplish-yellow, stalked, erect and appear with the leaves.

 FLOWERING TIME May–July.

HABITAT AND DISTRIBUTION It is native throughout Scandinavia (and also in the Alps) in tundra, rock ledges and other damp, open, calcareous habitats. In Britain, it only occurs in central Scotland, where it is uncommon.

Net-leaved Willow in flower.

Fuchsia
Fuchsia magellanica

A medium-sized shrub up to 3m tall, that is much-branched from the base, with arching stems. The bark is pale brown and peels off in strips. The leaves are ovate to elliptic, pointed, distantly toothed, short-stalked, up to 5cm long, in opposite pairs or threes. The flowers are conspicuous and distinctive, pendulous, on long slender stalks, with 4 bright red, large sepals up to 2.5cm long enclosing a ring of 4 purple petals. Protruding from the corolla are 8 long stamens, with a single, central, even longer style. The fruit is a blackish, ovoid or oblong, fleshy berry, though it is not always produced.

 FLOWERING TIME July–October.

HABITAT AND DISTRIBUTION A native of the temperate parts of South America, it is widely planted in parks and gardens for ornament and is used as a hedging plant in milder coastal areas, where it often becomes naturalised.

SIMILAR SPECIES There are many cultivars and varieties, with complicated nomenclature, which are planted for garden use and occasionally naturalised. Perhaps the most distinctive is Large-flowered Fuchsia *F. 'Corallina'*, which has a more spreading habitat, rather larger leaves and flowers, and a slightly inflated corolla. It is occasionally naturalised in milder coastal areas.

Left: Fushsia flowers.

Stag's Horn Sumach
Rhus typhina

A shrub or small tree, normally up to about 5m tall and occasionally taller, and often with multiple trunks, spreading branches and densely red-hairy twigs. The leaves are alternate, up to 12cm long and pinnately divided into up to 15 pairs of leaflets plus a terminal one. They are green in summer but turn a beautiful red colour in autumn. The male and female flowers are on separate bushes; they are both tiny and borne in dense terminal conical heads up to 15cm long. The female flowers develop into clusters of little nuts, each about 5mm across.

 FLOWERING TIME June–July.

HABITAT AND DISTRIBUTION A native of eastern North America, in the Cashew Nut family, that was introduced into Europe in the 17th century and is now widely planted for ornament. Although it rarely sets seed in northern Europe, it spreads quickly by suckering and is naturalised in the warmer parts of the area.

SIMILAR SPECIES Tanner's Sumach *R. coriaria* is a southern European species with a winged leafstalk and smaller flowers and leaves. It is occasionally planted.

Left: Stag's Horn Sumach, male flowers.

Smoke-tree
Cotinus coggygria

A spreading shrub with numerous branches from the base, that is up to 7m high, with hairless twigs. The leaves are ovate, often almost round, up to 8cm long, untoothed, waxy grey-green, often tinged pinkish and noted for turning a lovely, rich, redddish-pink in autumn before falling. The individual flowers are small, up to 1cm across, with 5 pale yellow petals aggregated into large diffuse inflorescences up to 20cm long. Most of the flowers fail to develop into fruit, growing instead into long, feathery, pinkish-purple plumes surrounding those few fruits that do develop – this is the 'smoke' of the common name.

 **FLOWERING TIME** May–July.

HABITAT AND DISTRIBUTION It is a native of warm, sunny habitats, often limestone, in south Europe from France eastwards. It is frequently grown in gardens – including as a variety of cultivars – and occasionally naturalised on roadsides and railway banks.

SIMILAR SPECIES An American species, *C. obovatus*, is often taller, with larger leaves but smaller inflorescences, and is grown for its intense autumn colour.

Smoke-tree foliage in autumn.

Smoke-tree in flower.

Horse-chestnut
Aesculus hippocastanum

A large, broad-domed tree, up to 25m tall (occasionally up to 40m). The bark is greyish or reddish brown, cracking or scaling to form large squarish plates. The twigs are distinctive – reddish-brown with numerous pale lenticels, terminating in large, glossy, brown, sticky buds up to 3.5cm long. The large leaves are palmately lobed and divided right to the base into 5–7 roughly oval, toothed, pointed leaflets up to 25cm long. The flowers are held in conspicuous, erect, pyramidal or cylindrical inflorescences up to 30cm long, made up of 40 or more white, 4- or 5-petalled flowers, with a red or yellow blotch at the base of each petal. The fruits are equally distinctive, having a large, spherical, spiny, green husk up to 6cm across, enclosing one or more glossy brown seeds – the familiar 'conker' – each with a pale scar.

 FLOWERING TIME May–June.

Fallen Horse-chestnut fruit in autumn.

HABITAT AND DISTRIBUTION

It is a native of mountains in south-east Europe, but widely planted almost everywhere, except in the far north, for ornament or shade. It is currently heavily affected by a bleeding canker disease.

SIMILAR SPECIES Red Horse-chestnut *A. x carnea* is smaller in all parts, with red or pink flowers and smaller, spineless fruits. It is of hybrid origin, commonly planted, and very occasionally naturalised.

Left: Horse-chestnut tree in flower.

Right: Horse-chestnut flowers.

Field Maple
Acer campestre

A small to medium, deciduous tree, up to 25m high, with a domed crown and sinuous trunk covered with brownish-grey, fissured bark. The twigs are paler brown. The leaves are up to 8m long and palmately lobed (like most maples) into 3–5 lobes, each with shallow rounded teeth. They are fresh green when young and turn bright yellow in autumn. The flowers are small, 5–6mm across, greenish yellow, with 5 small sepals and petals borne in small, axillary, erect clusters of up to 10 or so flowers, which appear at the same time as the leaves. Each fruit consists of two joined, winged seeds – known as samaras – with the wings held horizontally to each other. The winged seeds spiral gently downwards when detached, allowing new trees to germinate away from the parent.

FLOWERING TIME May–June.

HABITAT AND DISTRIBUTION It is native throughout much of Europe as far north as southern Sweden and northern England. It is planted elsewhere, most frequently on calcareous soils, in hedges, woodlands and scrub, and rarely becomes dominant.

Field Maple flowers.

Montpelier Maple leaves and flowers.

SIMILAR SPECIES Montpelier Maple *A. monspessulanus* is smaller, with simpler, 3-lobed leaves that turn red in autumn, and fruits in which the wings are held roughly parallel to each other. It is native as far north as northern France and southern Germany and is occasionally planted elsewhere.

Field Maple fruits.

Sycamore
Acer pseudoplatanus

A large, spreading, fast-growing, deciduous tree, up to 30m high, with a broad, domed crown. The bark is greyish-brown, mostly smooth, though cracking and flaking when old. The leaves are up to 15cm long, palmately lobed, with usually 5 lobes cut about halfway to the centre and each lobe toothed. The flowers are yellowish-green, individually up to 8mm in diameter, and clustered into cylindrical, pendulous inflorescences up to 12cm long that appear as the leaves open. The fruits are paired, winged seeds held roughly at right angles to each other.

 FLOWERING TIME April–May.

HABITAT AND DISTRIBUTION It is a native of central and southern Europe as far north as Holland, Belgium and Germany, in forests, copses, riversides and hedges on most soil types, but very widely planted and naturalised elsewhere for timber and ornament. In all but the coldest areas, it seeds readily and spreads rapidly, often dominating woods in areas where it is not native. It is common throughout the UK as a non-native.

SIMILAR SPECIES It is most similar to Norway Maple, see p. 146.

Leaves and flowers of Sycamore.

Sycamore tree in autumn.

Norway Maple
Acer platanoides

A tall, spreading, deciduous tree, up to 30m high, with a large, domed crown. The trunk is relatively short and covered with pale grey bark that is smooth when young but becomes ridged in older trees. The leaves are large, up to 15cm long, bright green and palmately lobed into 5–7 sharply toothed lobes (much more sharply toothed than sycamore leaves) and divided about halfway to the centre of the leaf. The flowers are bright yellowish-green, each about 7–8mm across, 5-petalled, in small, hairy, erect clusters that appear at the same time as the leaves. The fruits are the usual paired, winged seeds, with the wings held roughly horizontally to each other (compared to the roughly right angles in sycamore), and each one is up to 5cm long.

 **FLOWERING TIME** April–May.

HABITAT AND DISTRIBUTION It is a native of central and northern Europe as far north as southern Sweden and Norway, forests, copses, riversides and hedges on most soil types but also very widely planted and naturalised elsewhere for timber and ornament. It is absent from the UK as a native, but commonly planted and naturalised in warmer, lowland areas.

SIMILAR SPECIES It could be confused with Plane Trees (see p. 38), but differs in the flowers and fruit, and has quite different bark.

Above: Norway Maple leaves and flowers.

Above: Norway Maple autumn leaves.

Left: Norway Maple trees in autumn.

...leaved Lime

...latyphyllos

...ge but graceful tree up to 40m or so tall, with ascending ...nches that droop at the tips. The bark is dark grey and smooth ...n younger trees but becomes ridged and fissured with age. The twigs are reddish to olive-coloured. The leaves are broadly ovate to almost round, up to 10cm long, abruptly narrowed into a short, pointed tip, heart-shaped at the base and finely toothed on the margins. The undersides are paler and slightly hairy with whitish hairs. The flowers are greenish-white, fragrant and in clusters of up to 6 in a pendulous, branched inflorescence, whose common stalk is fused to a long, thin, pale green, papery bract, which remains attached to the fruits when they disperse. The fruits are ovoid to globose, up to 1cm long, hairy and 5-ribbed. The flowers are insect-pollinated and highly attractive to bees.

 FLOWERING TIME June.

Large-leaved Lime tree in autumn.

HABITAT AND DISTRIBUTION It is a native of central and southern Europe as far north as southern Britain, Denmark and south Sweden in woodlands and cliffs, mainly on calcareous soils. It is also widely planted and occasionally naturalised outside its natural range.

SIMILAR SPECIES Silver Lime *T. tomentosa* is similar in form but has silvery undersides to the leaves. Native to eastern Europe and west Asia, it is widely planted, especially in cities.

Right: Large-leaved Lime in flower.

Small-leaved Lime
Tilia cordata

A similar tree in outline and general appearance to Large-leaved Lime, up to about 35m high, with smooth greyish bark becoming fissured and cracked with age. The leaves are very similar in shape, but generally smaller, mostly less than 6cm long, though they may be larger on young growth. Underneath, the leaves are paler, with tufts of rusty brown hairs in the axils of the veins (which are absent from Large-leaved Lime). The flowers are similar in form and similarly attached to a narrowly elliptical bract, but they are held in an obliquely erect inflorescence, unlike the pendulous flowers of Large-leaved Lime. The fruit is ovoid, about 6mm long, downy at first, and obscurely ribbed.

 FLOWERING TIME June.

HABITAT AND DISTRIBUTION It is a native of central and southern Europe, except for the far north, in woodlands and cliffs, mainly on calcareous or heavy soils. It is also widely planted and occasionally naturalised outside its natural range. In Britain, it is mainly southern as a native, and considered to be a good indicator of ancient woodland. It was formerly coppiced, and coppice stools may be extremely old.

Left: Small-leaved Lime flowers and bracts.

Right: Small-leaved Lime tree in autumn.

Lime
Tilia x europaea

A tall tree, up to about 45m tall, and rather similar in outline to other limes but notable in having a trunk that is covered with outgrowths and epicormic shoots – i.e. sprouting shoots in clusters that do not develop into main branches. The leaves are similar in size and shape to those of Large-leaved Lime, with clusters of white hairs in the axils of the veins below, and more prominent veins above. The flowers are similar to Large-leaved Lime, borne in pendent clusters, which produce ovoid, downy, thick, slightly ribbed fruits that are often sterile. It is a hybrid between Large-leaved Lime and Small-leaved Lime that is both naturally occurring (albeit rare) and cultivated.

Above: Hybrid Lime trees with Mistletoe at Magdalen College, Oxford.

Right: Base of Hybrid Lime showing epicormic shoots.

FLOWERING TIME June.

HABITAT AND DISTRIBUTION It may occur naturally but uncommonly wherever the two parents are present, but it is much more likely to be seen as an introduced tree in parks, gardens, avenues and cities, occasionally becoming naturalised. It is less commonly planted in streets nowadays, thanks to the rain of honeydew from infesting aphids that are almost always present.

Mistletoe
Viscum album

Although not often thought of as a shrub, Mistletoe is actually a small shrub that grows on trees rather than on the ground! It is semi-parasitic (i.e. it is a parasite but also produces some of its own food from green leaves), yellowish-green and forms bushes up to about 2m across. The leaves are leathery, narrowly ovate, up to 8cm long, untoothed and in opposite pairs. The flowers are unisexual, 4-parted, on separate plants and in small, inconspicuous, greenish-yellow clusters in the leaf axils. The berries, on female plants, are conspicuous and familiar – globular, translucent white, fleshy, up to about 1cm across, and mainly ripening during the winter.

 FLOWERING TIME February–March.

HABITAT AND DISTRIBUTION It is widespread almost throughout Europe, as far north as southern Sweden, on a variety of deciduous trees, especially apple, poplar and others, or on conifers as a different subspecies (not in the UK). It is also widely used in festivals and is the subject of many legends.

Tree full of Mistletoe.

Mistletoe male flowers and berries (on separate plants).

Mistletoe berries.

Tamarisk
Tamarix gallica

A bushy, hairless shrub or small tree, usually up to about 3m, though occasionally much more in good conditions. The bark is purplish-brown and ridged. The leaves are very small, greenish-blue, scale-like, 1–3mm long and closely overlapping on the younger shoots. The individual flowers are minute, about 2mm across, with 5 pink petals clustered into thin, cylindrical racemes up to 5cm long. These racemes are clustered together at the tips of the branches to form larger, more conspicuous masses of flowers. The fruit is a small capsule, splitting to release seeds with an attached tuft of hair, which helps them to disperse. Tamarisk species are adapted to life on the coast, with an ability to secrete excess salt via special glands on the leaves.

 FLOWERING TIME June–September.

HABITAT AND DISTRIBUTION It is a native of coastal areas of south-west Europe on dunes, cliffs and the upper parts of saltmarshes. It is widely planted and naturalised elsewhere, especially on coasts. It is not native in Britain.

SIMILAR SPECIES African Tamarisk *T. africana* has larger, white flowers in larger racemes, and blacker bark. It is a native of southern Europe and often planted further north in similar habitats but rarely naturalised.

Right: Tamarisk tree growing on the beach.

Tamarisk flowers and leaves.

Dogwood
Cornus sanguinea

A deciduous shrub or very small tree up to about 4m high that is usually branched from the base but occasionally with a short narrow trunk. Older bark is greyish-brown, but the twigs are covered with shiny, reddish bark, especially noticeable in winter. The leaves are elliptical to oval, pointed, untoothed, up to 8cm long, with 3–4 pairs of main veins. If a leaf is carefully snapped and the two halves pulled apart, the glutinous latex from the veins remains attached to both halves of the leaf. The flowers are dull creamy-white, each about 1cm across, and borne in tight, terminal, erect clusters up to 5cm across; these are followed by clusters of globose black berries, up to 8mm across, and at the same time the leaves often go deep red before falling.

 FLOWERING TIME June–July.

HABITAT AND DISTRIBUTION It is native throughout much of Europe, including southern Britain, except the far north, mainly in scrub, woodland edges and hedgerows, especially on calcareous soils. It is planted and occasionally naturalised elsewhere outside its natural range, such as in Scotland and Finland.

SIMILAR SPECIES *C. sericea* from North America has blood-red twigs, narrower leaves and creamy-white berries, and it is often planted.

Dogwood in flower.

Dogwood berries in autumn.

Cornelian Cherry or European Cornel
Cornus mas

A small, spreading shrub or small tree, up to 8m high, with yellowish-grey twigs. The leaves are rather similar to those of Dogwood, ovate to elliptic, pointed, untoothed and up to 10cm long. The flowers are small, yellowish, 4-parted, about 5mm across and grouped into umbellate clusters of 20–25 flowers, with 4 petal-like bracts at the base of the umbel that appear conspicuously in early spring before the leaves. The fruit is a distinctive bright, shiny, red, oblong-ovoid berry, up to 2cm long, on a short pendulous stalk, fleshy and rather acidic – hence the name Cornelian Cherry, though they are not related to true cherries. In parts of eastern Europe, they are valued for making jam and chutney, as well as alcoholic drinks, and used when they are very ripe.

FLOWERING TIME February–March.

HABITAT AND DISTRIBUTION It is a native of central and eastern Europe as far north as France, Belgium and south Germany, but widely planted for its attractive late winter flowers and its fruit. It is occasionally naturalised, e.g. in Britain.

SIMILAR SPECIES Horticultural varieties with variegated leaves, yellow fruit and others may occasionally be found.

Cornelian Cherry flowers.

Cornelian Cherry bush in spring.

Strawberry Tree
Arbutus unedo

An evergreen shrub or small tree, up to 9m high and occasionally more, with a dense, rounded crown and a short trunk. The bark is reddish-brown at first, then peeling into short, hanging strips. The young twigs are attractively pinkish or red. The leaves are narrowly ovate, up to 10cm long, dark green, pointed, finely toothed or untoothed, with a prominent central midrib. The flowers are white, sometimes tinged with pink or green, bell-shaped, fragrant, each up to 1cm long and clustered in a dense, drooping panicle up to 6cm long. The fruits are distinctive – a spherical, warty berry, up to 2cm in diameter, which ripens from yellow to rich red. Because the fruit takes about a year to mature, there are normally ripe fruits and flowers on the tree at the same time, in autumn. The fruits are edible, but not especially nice.

 FLOWERING TIME September–December.

HABITAT AND DISTRIBUTION It is a native of the Mediterranean region, where it occurs in woodland, scrub and rocky places, with an anomalous outpost in south-western Ireland, though the native status of these Irish trees is now uncertain. It is also widely planted outside its range, and naturalised in acid or calcareous areas here and there.

The distinctive mature bark.

Flowers and a single fruit.

Rhododendron
Rhododendron ponticum

A strong-growing, dense, evergreen shrub, with spreading to ascending branches, up to 5m high. The bark is reddish on young growth and becomes greyer and more fissured on older wood. All parts of the plant are hairless. The leaves are elliptic to oblong, leathery, up to 20cm long, untoothed, dark, shiny and green above but paler below. The flowers are large, mauvish-purple, darker in bud, up to 6cm long, funnel-shaped or trumpet-shaped, and borne in terminal clusters of up to 15 flowers.

 FLOWERING TIME May–July.

Rhododendron in flower.

HABITAT AND DISTRIBUTION The species is native to scattered parts of south-west Europe and west Asia, though it has been widely planted, and abundantly naturalised, in the milder parts of northern Europe, such as Ireland and the UK, in a wide variety of habitats on more acid soils. Recent analysis has shown that the naturalised plants are mainly the Iberian subspecies, ssp. *baeticum*, though there is also much hybridisation.

SIMILAR SPECIES Many other Rhododendron species and hybrids are planted in gardens and parks. Alpenrose *R. ferrugineum* is native in the Alps and planted elsewhere.

Alpenrose in flower.

Labrador-tea
Rhododendron tomentosum

A small, evergreen shrub, up to 1.5m tall, erect or spreading, with rusty-hairy young twigs. It was formerly known as *Ledum palustre* but has recently been reclassified as a *Rhododendron*. The leaves are narrowly oblong, dark green above, untoothed, up to 5cm long, with rolled-over margins and a dense covering of rusty hairs below. The flowers are starry, creamy-white, 5-petalled, up to 15mm across, normally with 10 stamens, and grouped into dense, attractive, terminal clusters. The fruit is a dry, inconspicuous capsule.

 FLOWERING TIME May–July.

HABITAT AND DISTRIBUTION It is native from northern Germany northwards throughout most of Scandinavia in damp, acidic habitats such as bogs, coniferous woods and moorland. It is also often planted, and occasionally naturalised. Planted specimens are frequently the North American species *R. groenlandicum*, which has narrower leaves with a concealed midrib on the underside and only 8 stamens, but there is a good deal of confusion between the species. Plants naturalised in Britain are all considered to be the North American species.

Labrador-tea in flower.

Labrador-tea wild in Scandinavia

Tree Heather
Erica arborea

A medium-sized, bushy shrub, up to 4m high, with erect branches. The leaves are tiny, 5–7mm long, narrowly linear, in whorls of 3 or 4, bright green and hairless. The flowers are pure white, narrowly bell-shaped, 3–4mm long, and clustered into lateral racemes collectively forming a conspicuous terminal head. The stigma in this species is white.

FLOWERING TIME March–May.

HABITAT AND DISTRIBUTION It is a native of southern Europe but widely planted for ornament, and occasionally naturalised in mild regions on acid soil, such as south-west England, where it occurs along wood margins and on heathland.

SIMILAR SPECIES
Portuguese Heath, or Portuguese Heather *E. lusitanica*, is very similar in general appearance, but the flowers are more often tinged pink, slightly larger (the corolla is 4–5mm long) and have a red stigma. A native of acid soils in south-west Europe, it is often planted for ornament and occasionally naturalised in heathy places in the milder regions. Irish Heath *E. erigena* has pinker flowers with protruding reddish anthers. It is native to western Ireland and south-west France, and occasionally planted elsewhere.

Tree Heather in flower.

Portuguese Heath in flower.

Duke of Argyll's Teaplant
Lycium barbarum

A deciduous, intricately branched, slightly spiny, spreading shrub, up to 3m tall. The leaves are narrowly elliptic, widest around the middle, up to 10cm long, untoothed and greyish-green. The flowers are 5-petalled, purplish becoming brown, trumpet-shaped, about 1cm across, with protruding stamens, borne singly or in very small clusters. The fruit is a bright red, roughly ovoid berry, up to 2cm long.

 FLOWERING TIME June–September.

HABITAT AND DISTRIBUTION It is native to China but widely planted, especially in coastal regions, and naturalised as far north as southern Scandinavia. The curious English name comes from an attempt to introduce tea plants from China to Britain in the early 18th century, but a mix-up of labels led to the wrong species being planted!

SIMILAR SPECIES Chinese Teaplant *L. chinense* has broader leaves, widest below the middle, and larger flowers. Also from China, it is planted and naturalised over a similar area, but less commonly. It is the source of the commercially available 'Ghoji berries'.

Ripe Ghoji berries from Chinese Teaplant.

Duke of Argylll's Teaplant in flower.

Privet
Ligustrum vulgare

An erect, semi-deciduous shrub, up to about 4m high. Its bark is greyish or greyish-brown and smooth. The young twigs are hairy at first but soon become hairless. The leaves are lanceolate, up to 4cm long, pointed, untoothed, short-stalked and persistent. The flowers are white, made up of a short tube and 4 spreading petal lobes, about 4–6mm across, and gathered into dense, pyramidal, terminal inflorescences. It is fragrant, but this is not usually considered to be pleasant. The fruit is a globular, shiny, black berry, up to 8mm across, produced in abundance.

 FLOWERING TIME May–June.

HABITAT AND DISTRIBUTION It is widespread throughout much of Europe, northwards as far as southern Sweden, in sunny and semi-shaded situations such as wood margins, hedgerows and scrub, particularly on calcareous soils. In Britain, it is native in England and Wales, but not native in Scotland or Ireland.

SIMILAR SPECIES Garden Privet *L. ovalifolium* has more fully evergreen and broader leaves (though it may lose them all at once in very cold winters) and slightly larger flowers. It is native to Japan, but widely planted and occasionally naturalised.

Privet berries.

172

Privet in flower.

Ash

Fraxinus excelsior

A large tree, up to 40m tall, with thick twigs, a high, open crown, and smooth, grey bark that becomes cracked and fissured in older trees. The twigs are thick, grey, flattened at the nodes and terminate in large, conical, sooty-black buds. The leaves are in opposite pairs, pinnate, up to 35cm long, and divided into 3–6 pairs of oval leaflets, together with a terminal one. The flowers are very small, purplish, and in axillary or terminal clusters appearing before the leaves. The flowers may be unisexual or hermaphrodite and – if unisexual – males and females may be on separate trees or not. The fruits are distinctive 'keys' – nuts with a single wing, narrowly ovate, up to 5cm long and green at first but becoming papery brown.

 **FLOWERING TIME** April–May.

HABITAT AND DISTRIBUTION It is widespread throughout much of Europe, except the far north, in woods, scrub and hedges on a wide variety of soils from dry limestone to heavy clay. At present, a highly infectious and damaging disease, Ash Dieback, is spreading rapidly through Europe.

SIMILAR SPECIES Narrow-leaved Ash *F. angustifolia* differs in having narrower leaflets that are more coarsely toothed and brown winter buds. A native of southern Europe, it is planted as a street tree and ornamental.

Above: Ash buds just opening.

Left: Ancient Ash tree in summer.

Holly
Ilex aquifolium

An evergreen shrub or small tree, normally up to about 10m high, but specimens up to 23m have been recorded. The bark is silvery grey and smooth at first but becomes fissured and warty with age. The leaves are distinctive, dark, leathery, green, ovate to elliptic in general shape, but with strongly wavy margins and very distinct spines, although the leaves vary and upper branches often bear much less spiny leaves (presumably because they are out of reach of browsing animals). The flowers are white, small (4–6mm across), 4-petalled, in small clusters and unisexual, with male and female flowers found on different trees. The fruit are the familiar globular, bright red berries, about 1cm across, borne in clusters and widely used at Christmas. Of course, only female trees produce berries.

 FLOWERING TIME May–June.

Above: Holly berries.

HABITAT AND DISTRIBUTION
It is widespread throughout much of Europe, northwards as far as northern Scotland and southern Scandinavia, in woodlands, clearings, hedges, scrub, gardens and rocky places. It is resistant to grazing, so it does well in rough pasture.

SIMILAR SPECIES There are no other native species, but Highclere Holly *I. x altaclerensis* – a hybrid of garden origin – is widely planted. It has larger, flatter, less spiny leaves, and larger fruits and flowers.

Left: Holly in flower on limestone pavement.

Common Elder
Sambucus nigra

A deciduous shrub or small tree up to 10m high, with corky, uneven, grey-brown bark. The twigs are curved, brittle and have a central soft pith. The leaves are up to 15cm long, pinnately divided into 5–7 narrowly elliptical, toothed, pointed leaflets. The flowers are small, up to 8mm across, creamy-white and clustered into large, conspicuous, flat-topped, scented clusters up to 25cm across. The berries are globose, each up to 6mm across, ripening to dark purplish-black and produced in great quantities. The fruits are widely used for wine and preserves, and are highly attractive to birds. The flowers are often used to make both alcoholic and non-alcoholic drinks.

 FLOWERING TIME June–July.

HABITAT AND DISTRIBUTION Widespread throughout northern Europe, except the far north, in sunny and semi-shaded, nutrient-rich situations such as farmyards, around buildings, roadsides and hedges, particularly on calcareous soils. In Britain, it is native throughout Britain and Ireland except for the northern islands. It is often planted for ornament and its fruit.

Elder flowers and leaves.

Ripe elderberries in late summer.

Elder bush in flower.

Red-berried Elder or Alpine Elder
Sambucus racemosa

A deciduous shrub, up to 4m high, similar in general appearance to Common Elder, though usually smaller. The leaves are pinnate, up to 20cm long, with 3–7 leaflets, most commonly 5. The flowers are creamy-white or yellowish, small, clustered into dense ovoid to pyramidal, terminal panicles (not flat-topped like Common Elder) up to 7cm long. The berries are globose, ripening to bright scarlet-red.

 FLOWERING TIME April–July.

HABITAT AND DISTRIBUTION It is native to the Alps and other mountain areas, but widely planted and naturalised as far north as southern Scandinavia. It is commonly naturalised in northern Britain.

Dwarf Elder in flower.

SIMILAR SPECIES Dwarf Elder or Danewort *Sambucus ebulus* is a spreading, rhizomatous, rather shrub-like herbaceous plant up to 2m tall, without woody stems. The leaves are similar, but have large, persistent stipules at the base. It has flat-topped heads of dull white or pinkish flowers, producing small, black berries, though neither as shining nor luscious as those of common elder. It is native to parts of southern and western Europe, but widespread as an ancient introduction further north, including most of Britain, and northwards in northern Europe as far as southern Sweden.

Red-berried Elder berries.

Red-berried Elder flowers.

Guelder-rose
Viburnum opulus

A small, spreading, much-branched, deciduous shrub or very small tree up to 5m high, with smooth, angled, greyish-brown twigs and wavy branches, and with buds that have scales. The leaves are palmately lobed, up to 8cm long, with 3–5 irregularly toothed lobes and thread-like stipules at the base. The flowers are creamy-white and gathered together in large, flat, plate-like inflorescences up to 12cm across. The outermost flowers are large and sterile, borne in a ring surrounding a mass of smaller (up to 7mm across) fertile flowers in the centre, making the whole look like a single, large flower. The fruit is a rounded, glistening, shiny, translucent red berry, up to 11mm long, hanging down in clusters. The plant is not related to true roses, nor does it resemble them.

 FLOWERING TIME June–July.

HABITAT AND DISTRIBUTION This is a widespread native throughout northern Europe, except northern Scotland and northern Scandinavia, in hedges, scrub, damp woodland and other habitats.

SIMILAR SPECIES There are garden cultivars with yellow berries, which are rarely naturalised. Some forms of the garden plant *Hydrangea paniculata* can look rather similar, but the leaves and fruits are quite different.

Guelder-rose in flower.

Guelder-rose berries and autumn leaves.

Wayfaring-tree
Viburnum lantana

A downy, deciduous shrub or very small tree up to 6m high. The twigs are greyish and hairy, and the buds have no scales. The leaves are oval to lanceolate, finely toothed, hairy, and white-woolly underneath. The flowers are creamy-white, small, 5–9mm across and borne in dense clusters up to about 10cm across, lacking the sterile ring of flowers that occurs in Guelder-rose. The fruits are flattened, ovoid berries, becoming red then finally black when ripe, and all ripening at a different rate so the head is often a mixture of all colours.

 FLOWERING TIME April–June.

Wayfaring-tree berries.

Wayfaring-tree in flower.

HABITAT AND DISTRIBUTION It is native in Europe as far north as Germany, Belgium and northern England in calcareous, open habitats such as banks, hedges, scrub and dry grassland. It is naturalised in northern Britain and as far north as southern Sweden and Norway.

SIMILAR SPECIES Laurestinus *V. tinus* is closely related, though it differs in being a taller, evergreen bush, with darker, glossy leaves. The flowers are pink outside, white inside, in clusters, and produce bluish-black berries. It is a native of scrub in southern Europe, but widely planted in parks and gardens for its foliage and early flowers, and becoming naturalised in warmer regions.

GLOSSARY

anthers: the pollen-bearing part of a stamen.

auricle: a lobed extension of the lower part of the leaf blade, particularly significant in the oaks.

axillary: something found in the axil (i.e. junction) of a twig with a main branch, or a side vein with a main vein.

bract: nodified leaf that supports a flower, often quite reduced and scale-like in nature.

calcareous: referring to the characteristic of a soil, and indicating that the soil has high levels of calcium, derived from limestone, chalk or similar rock types.

calyx: the outermost whorl of a flower, normally consisting of the green sepals that enclose the developing bud.

carr woodland: consistently wet woodland, such as around a lake or on a river floodplain.

catkin: a tightly packed spike of reduced flowers on a long axis, most commonly pendulous and flexible in nature. Usually associated with wind-pollinated flowers.

columnar: referring to tree-shape, indicating a tall and very narrow shape, such as a Lombardy Poplar (see p. 13).

cone: a compact and usually ovoid organ, made up of a collection of scales bearing seeds or spores; most associated with the conifers (pp. 12–34), but also applied to the fruits of alders.

conical: cone-shaped. When used in reference to tree shape, it indicates a tree that is broader at the base, tapering gradually to a narrow top.

coniferous: literally, cone-bearing, but used specifically to refer to a group of primitive trees with needle-like leaves and bare seeds that are not enclosed within a fleshy ovule. The coniferous trees are on pp. 12–34.

deciduous: refers to trees that shed their leaves every autumn, remain bare through the winter, and regrow them each spring.

dioecious: having the male and female flowers on separate plants. Although often similar in appearance, only the female plants produce fruit (e.g. Holly).

evergreen: refers to trees that retain their leaves all year round. The leaves are normally tough or leathery, to withstand the cold of winter as well as summer heat.

hermaphrodite: with male and female parts in the same flower.

inflorescence: a distinct cluster of flowers, which may be almost any shape, and includes catkins, spikes and umbels as examples.

lanceolate: usually refers to the shape of a leaf, indicating long and narrow, and roughly six times as long as it is wide (see p. 126).

larval food plant: a plant that is specifically fed on by the caterpillars of a moth or butterfly.

lenticels: small visible corky pores or lines on the surface of the stems of woody plants that allow the interchange of gases between the interior tissue and the surrounding air.

limestone pavement: a distinctive habitat, most commonly on Carboniferous limestone, where glaciation has left a flat limestone layer with plateaux intersected by cracks. A few plants do particularly well in this harsh environment.

midrib: the central vein of a leaf or bract.

monoecious: having separate male and female flowers, but on the same plant.

naturalised: a plant that is not native to an area, but which has established itself as part of the local vegetation.

ovate: the most common shape of a leaf, broadest at the base and roughly 2–3 times as long as wide.

palmate: a lobed leaf with all the lobes arising from the same point (see p. 142).

panicle: a branched inflorescence, with branches arising up the stem.

petiole: the stalk of a leaf; the broad flat part is known as the blade.

pinnate, pinnately divided: a lobed compound leaf (or other structure) where the lobes arise in pairs along the length of the axis (see p. 66).

pollard, pollarded: trees that have been cut off at about head height or above, with branches growing from the cut stump.

raceme: a spike-like inflorescence with stalked flowers, where the lowest are the oldest.

sepals: see calyx.

sessile: without a stalk.

stipules: small leaf-like or bract-like growths at the base of a leaf stalk (see p. 129).

style: the elongated stalk-like part of an ovary, bearing the stigma, onto which pollen falls.

ternate: a compound leaf with 3 leaflets.

unisexual: a plant or flower that is either male or female.

BIBLIOGRAPHY AND RESOURCES

There are many books dealing with various aspects of European trees and shrubs. Those listed below will lead you further into this extensive literature.

Milner, E., *Magnificent Trees of Britain: History, Folklore, Products and Ecology*, Anova, London, 2012. More information about trees and their relationship to man.

More, D. and White, J. *Trees of Britain and Northern Europe*, Bloomsbury Publishing, London, 2013. An encyclopaedic book on Europe's trees.

Stace, C., *New Flora of the British Isles*, third edition, Cambridge University Press, Cambridge, 2010. The most complete and up-to-date flora of the British Isles – including all the native and many introduced trees and shrubs.

ORGANISATIONS AND SOCIETIES

Below are the names and websites of various organisations and societies that promote an interest in trees, with a brief note on the particular interests of each society.

Botanical Society of Britain and Ireland (BSBI)
www.bsbi.org.uk
The main UK society for the study and recording of wild plants, including trees and shrubs.

Field Studies Council (FSC)
www.field-studies-council.org
An educational charity that runs courses, including many on plant identification, at a number of field centres around the UK.

Plantlife
www.plantlife.org.uk
The major UK and European charity that works towards the protection of wild plants.

Wildlife Trusts
www.wildlifetrusts.org (for UK)
http://iwt.ie/ (for Ireland)
The county wildlife trusts are conservation organisations covering
every UK county, working for the protection of species and habitats.

SUPPLIERS

NHBS
http://www.nhbs.com/
Primarily a supplier of a huge range of natural history books, but
now also supply hand lenses, binoculars and many other items of
fieldwork equipment.

Watkins & Doncaster
http://www.watdon.co.uk/the-naturalists/
Natural history suppliers, based in Kent.

PLACES TO VISIT

Bedgebury National Pinetum
Park Lane
Goudhurst, Kent
http://www.bedgeburypinetum.org.uk/
Has a huge collection of conifers.

Hillier's Arboretum
Sir Harold Hillier Gardens
Jermyns Lane
Romsey, Hampshire
http://www3.hants.gov.uk/hilliergardens
Has an excellent collections of trees.

Westonbirt Arboretum
The National Arboretum
Nr Tetbury, Gloucestershire GL8 8QS
http://www.forestry.gov.uk/westonbirt
Has an internationally renowned tree collection.

INDEX

INDEX